Ghostly Bedfordshire... reinvestigated

by

Damien O'Dell

To: the late Bill Turner & the people of Bedfordshire –
a county to take pride in.

ISBN 0 904160 76 X

Printed in England by Streets Printers, Royston Road, Baldock,
Herts SG7 6NW

Damien O'Dell's first book about the paranormal brings together his various interests – his Bedfordshire roots, his fascination with the paranormal and his love of history. He travelled many hundreds of miles, from Odell to Arlesey, from Barton-le-Clay to Bolnhurst, and interviewed scores of witnesses in his search for the truth. His book is filled with a rich panoply of characters, both famous and infamous, from Sir Fulke de Breaute to Lord Thomson of Cardington. Well over a hundred people feature in Ghostly Bedfordshire…reinvestigated, but the most interesting stories come from 'ordinary' people, from a variety of backgrounds. Their sincerity, often corroborated, where possible by other witnesses, convinced the author that what we refer to as 'ghosts' are very real experiences. This was reinforced in the village of Wilden, when he came into spine-tingling, personal contact with poltergeists. What follows has been written with professional thoroughness, it may seem incredible, even unbelievable…but it is absolutely true…welcome to Ghostly Bedfordshire…reinvestigated – a haunted history.

CONTENTS

ACKNOWLEDGEMENTS

Special thanks go to the many friends without whom this book would
not have been possible:

Paul Bowes

Jennie Butterworth

Gill Campbell

Jennie Clarke

Michael Cook

John Duggan

Keith & June Paull

Simon Peters

Christopher Robinson

Tony Toth

And of course my very best friend and my 'organiser', my wife:
Vicki

Foreword to the New Edition

As a collector of ghost stories since childhood, I have eagerly devoured scores of collections, such as Lord Halifax's Ghost Book and Elliott O'Donnell's Haunted Britain, also I have keenly followed the adventures of ghost-hunter Harry Price, and been enthralled by his The Most Haunted House In England, which chronicles his investigations at Borley Rectory, allegedly England's, if not Britain's, most haunted house. Now older and, hopefully, wiser, I am not so sure about Borley's reputation, although it was certainly haunted; the claims for it seem exaggerated in the light of the latest evidence. Harry Price most definitely advanced the study of parapsychology and his enduring legacy is his library – The Harry Price Collection, over 15,000 volumes now housed in the University of London Library.

My own initial encounter with the paranormal occurred in 1962, at the age of twelve, when my family lived just off the North End Road, in London's West Kensington. Immediately behind our house was an overgrown estate, known as The Grange. The original house had long since fallen into ruin; its ultimate fate was to be redeveloped into council flats. In the long distant past many famous visitors had come to The Grange, including the writer Rudyard Kipling, Dante Gabriel Rossetti, poet and painter, and William Morris, craftsman, poet and political activist, but I was unaware of that then. It wasn't until many years had passed that the whole story fell into place. On a visit to Wimpole Hall in Cambridgeshire, I learned, from an exhibition about Kipling, that the grand house that once stood behind our modest home in Matheson Road belonged to the illustrious Romantic painter and designer Sir Edward Coley Burne-Jones (1833-1898). One night, as my best friend, Renato Bondonno, and I sat up looking out of my bedroom window, he nudged me, and in a fear-choked voice said, 'Do you see that?' I saw nothing for a while and then as my eyes adjusted to the moonlit night I could see what he saw; a figure, roughly man-shaped and all in white. As it glided by in the gardens I could clearly see the trees through it, the figure was transparent. Where Renato was somewhat agitated I was, instead, fascinated, for I was certain that what we had seen was a ghost. No other explanation would fit, and like any young boy, I was excited. Was it the shade of Burne-Jones himself, who had died some sixty-three years earlier at the age of sixty-five? I like to think that it was.

I suppose that it was inevitable that I should write about ghosts in general, and ghosts of Bedfordshire in particular, in later life. Although born and brought up in London, a city I love, my forbears originate in Bedfordshire. All of us with the surname O'Dell, in its several variants (Odell, O'Dell, O'dell), owe our ancestry to a small village in North Bedfordshire (to the north west of Bedford), called Odell. The Alston family lived in the village for 300 years, and they are commemorated in the local church; their ancestral home is on the site of ancient Odell Castle. There is a legend concerning Sir Rowland Alston, a wicked man, whose ghost was alleged to terrorise the district, walking through walls and trees in broad daylight, and riding his black horse into the hall of the ancestral home. Twelve clergymen conducted an exorcism, and successfully consigned the restless spirit to a pond on Odell Wold. After a hundred years Sir Rowland emerged from the pond, but was chased by the Devil. Sir Rowland's wraith squeezed through the keyhole of the church, however, and the furious Devil shook the church, leaving five giant fingermarks on the stone jamb of the porch. Sir Rowland's haunting became reduced to once every 100 years; he is due to ride around his former estate in 2044…

A few years ago my work took me to Bedfordshire, first to Biggleswade, and then to Bedford. In the course of my job I came across several, ordinary down-to-earth people, who had had brushes with the supernatural. They related them to me in a matter-of-fact way, but they were so interesting that the idea for this book was born. I have endeavoured to maintain that interest in the stories: to keep them as contemporary as possible, to interview first-hand witnesses for preference, to illustrate these accounts with photographs that would set the scene, to research some of the history of these haunted buildings, and to investigate some of the immediate cases personally, and with friends, in order to amass as much evidence as possible. What follows is true; I hope you enjoy reading it even half as much as I have enjoyed researching and experiencing it…

Since the exciting times when I successfully launched my first book, it became apparent that a second edition was necessary, to bring readers up to date with many important developments. One of these was meeting Roger Ward, Chairman of the Friends of Chicksands Priory, and champion of that unique building. Roger is mainly responsible for protecting and promoting this most important (and most haunted) piece of English history. I greatly enjoyed his guided tour of the Priory, and you too, can participate, by contacting the Tour Co-ordinator on 01525 860497. Roger contacted me after reading a copy of the first edition and he was able to supply some new and absorbing facts.

During the series of talks that I gave, while promoting the book, I met many members of the public who share my enthusiasm for the paranormal. Luton Library was a particularly memorable visit, on the chilly night of March 23rd this year, when 150 people packed into the building, filling every available seat in the place! I'd like to thank you all for your enthusiasm – and look out for Ghostly Hertfordshire, which will be available from all good bookshops next year... Certainly meeting my readers brought forth yet more great new stories and all of them appear in this edition, including several pub ghosts, like the Moggerhanger inn haunted by the spirit of a little boy. I also kept abreast of the latest paranormal developments at the Cecil Higgins Art Gallery, was invited into a haunted radio station in Luton, uncovered some facts concerning Clophill's notorious church, researched three new chapters on Biggleswade, Leighton Buzzard and Dunstable, and you will find the first public account of the spooky Dunstable Priory House investigation... but hey, read for yourself, and enjoy a bit of spine-tingling!

Damien O'Dell
October 2004

Chapter 1

Dunstable Dreams and Visions

The Dreams...

Dunstable was originally known as Dunstaple, an Anglo-Saxon name from 'Dun' (a hill) and 'Staple' (a market), hence 'The market by the Downs'. The area was populated long before this however, prehistoric burial mounds and earthworks on the chalk hills surrounding the town clearly attest to its early importance. It was formerly a Roman posting station on the crossroads of prehistoric Icknield Way and Roman-built Watling Street. The site was abandoned in Saxon times, but rose to prominence some time before 1109 when it was established by Henry I, son of William the Conqueror, who later had his 'palace' of Kingsburie built here. A Royal Hunting Lodge was also created, which accommodated his court whenever the King visited the area. Built in the shadow of St Peter's Church, parts of it can be seen today, including a fine Norman nave and magnificent west front. The Royal Hunting Lodge site still exists, now in use as a hotel and a nearby public house, The Norman King. Henry regularised the markets held at Watling Street and Icknield Way crossroads and granted local control over them to the Priory of St Peter in the Town Charter of 1131. Dunstable's fame was spread by its association with Queen Eleanor, Edward I, Henry VIII and Elizabeth I, as we shall see later... The Reformation saw the dissolution of the Priory of St Peter and the town recovered from this setback to gain importance during the advent of stagecoach travel and several coaching inns remain from this prosperous period, which itself declined with the coming of the railway age and the consequent shift of the straw plait and hat-making industries from Dunstable to Luton, which had a railway station. The town adapted again, at the turn of the century it welcomed new industries such as printing, engineering and the motor trade.

Dunstable undoubtedly has a major claim to paranormal fame, it is home to the most remarkable psychic that I have ever encountered – possibly the world's most highly skilled exponent of precognitive dreaming – Chris Robinson. The renowned 'psychic detective' lives with his wife Bessie and their children in an unremarkable, modest terraced home, not far from High Street South. I had the good fortune to meet Chris when he agreed to give a talk, on Sunday, the 12th September 2004, at the club which I founded, Anglia Paranormal Investigation Society (APIS), at The Five Bells pub, the group's regular meeting place at Cople, near Bedford. Chris had recently returned after months of scientific testing under the supervision of Dr Gary Schwartz, at the University of Arizona. The tests were overwhelmingly positive in confirming Chris's ability to somehow 'dream' future events. Chris has worked with British Intelligence, HM Customs and Excise and the Metropolitan Police Force in dealing with terrorist threats and criminal activities. Absolutely fascinated by his talk, during which he was shadowed by a Japanese television crew making a programme about him, I quickly set up an interview of my own with the 'dream detective' as he is known, for the following Tuesday morning.

It was surprising that I hadn't heard of Chris before his talk, because he showed me a bulky file of press cuttings and articles about his gift, and I also left his home with a compilation of his various television appearances, all of which I had managed to miss in the past. While I was interviewing him I was struck by his down to earth nature, his 'ordinariness', Chris earns his living as an office cleaner and he used to be a television repairman. But he has a quite extraordinary ability – he has precognitive dreams which predict future events with uncanny accuracy and he has supplied valuable information to the security services, on numerous occasions. As long ago as 1999 Chris dreamt that Muslim terrorists were going to hijack airliners and crash them into western cities, and he reported this to his contact at British Intelligence.

Precognition is impossible according to science, and this has always made it difficult for law enforcement agencies to take psychics seriously. Scotland Yard already had evidence to the contrary, after years of information supplied by Chris they knew that he had a genuine gift. Despite that, what Chris was now proposing just seemed to stretch their credulity too far. In 2001 Chris decided to try and prove his abilities once and for all, so he contacted Dr Gary Schwartz of the Human Energy Systems Laboratory at the University of Arizona. Dr Schwartz agreed to help – he would set up strictly controlled scientific testing which would prove, one way or another, the validity of Chris's precognitive claims. While Chris was in Arizona, in August 2001, he dreamt of planes being crashed into the World Trade Centre in New York. Two days before 9 - 11 Chris wrote to the US Embassy in London in an attempt to warn them...

The Japanese television people, whom I met during Chris's talk to APIS, were from Coool Communications and they were currently working with Chris after setting him a challenge. Back in Tokyo there was a sealed envelope and they had asked the psychic to dream about what was in the envelope. In his dream Chris's late father visited him and said 'It's an Australian bird, with two heads.' Chris rapidly translated this – his father, when he was alive, had referred to women as 'birds' and Australia had been their code word for dead (down under). So Chris knew that the envelope contained the images of a dead girl, either two photographs of her, or two photographs of two separate dead girls, and this is what he told his enquirers. The Japanese were on the next flight to Luton, they wanted help, and the dead girl had been a murder victim.

On the Friday before our interview Chris had a dream visit from the dead Japanese girl and he was given a word that made no sense to him, as it was Japanese, but when he repeated it, as best he could, to the TV crew they recognised it immediately – it meant danger. He also told me about something that he had never experienced before, on waking from dreaming about four Japanese men he discovered that they were still there in the room with him, seemingly solid and real, before their figures slowly faded away. His instinct tells him that these are some kind of spirit guardians to Yoko (the murdered girl). He was given her name by her spirit, and the information that she had been strangled, at 9.45pm; these facts were also later verified. Some of the communication was in English and some in Japanese, but nearly everything Chris dreams has to be interpreted anyway, as most of it is symbolic, he works extremely hard at making sense of it all. He often gets huge amounts of information – 'You hope you remember the important stuff' he commented. I am glad that I don't share Chris's 'gift', it seems a huge burden, an enormous responsibility to me. His next task for the Japanese is to try and dream the name of Yoko's killer...

During the dream detective's presentation at Cople we saw the amazing video that he had made in America, while undergoing scientific testing by the University of Arizona in Tucson, organised and controlled by Professor Gary Schwartz. The most notable feature of the video is the gradual conversion of Professor Schwartz from outright sceptic to firm believer in Chris's precognitive powers. The test was simple – Chris was to be taken, by Professor Schwartz, to ten separate sites, chosen completely at random in this huge American State. Gary Schwartz wanted to rule out any kind of possible telepathy and he chose twenty different locations. The name of each place was typed on some paper which was sealed in an envelope. He then passed these twenty envelopes on to a magician friend, who lives in New York; he in turn gave them to a third party. Out of the original twenty envelopes ten were randomly chosen for the tests. On each day of the ten tests, after Chris and Dr Schwartz had recorded the dreams, the person holding the envelopes was instructed to open one. This was the site to visit that day. Chris knew none of the places, but his dreams proved to be accurate in predicting details of the various locations which they were due to visit. Although the psychic was correct in all of his descriptions of the sites, it was clear that the scientist was not convinced at first. Then – on day four – the results were so overwhelmingly in favour of precognition that Professor Schwartz was obliged to accept the only possible explanation for the veracity of Chris's predictions – he could, somehow, accurately predict future events through the power of his dreams.

In Chris's Dream Diary (since 1990 he has kept a written record of the impressions he receives while asleep) for the fourth day he recorded a dream where he was in a café in Greek Street, Soho. He had two rolled up newspapers – The Sun and The Mirror, which he used as a telescope to look at the sky and he also had a camera, he noted that it was an Olympus. He decoded this vision to mean that the two men would be visiting a mountain (Olympus) where they would see some kind of telescope, or mirror, used to look at the sun. It is clear on the film that this revelation completely blew Schwartz away, because on that Sunday, which was his birthday, he opened the sealed envelope to discover that they were to visit an observatory at Kitt Peak (Solar Observation Helioscope Observatory – SOHO) on the top of an Arizona mountain! All of the dream location predictions are amazing, but this one in particular stands out, and it is as if whoever, or whatever, is guiding Chris, has decided to pull out all the stops to convince the doubting scientist. The dream detective is, as usual, open about the guidance he receives – 'It could be alien intelligence, God-like intelligence, I just don't know' he told me.

I can only begin to imagine the frustration that Chris must feel at times when his warnings go unheeded. During our interview he told me about a case that featured in all the daily newspapers, back in 2002 two little girls, Holly Wells and Jessica Chapman went missing in Soham, Cambridgeshire. Chris had a dream, the night before they died, about the children playing in the snow; in his dream language snow is always synonymous with imminent danger and he feared the worst. The girls later returned to say that they were worried about animals eating their remains, thereby confirming the fact that they were now dead. In another dream he found himself walking down the runway at the air force base at Mildenhall, and looking at the perimeter fence. He dowsed a map of the area and pinpointed the location of their bodies. Chris tried to warn the Cambridgeshire police that the girls were dead, but was ignored, and the force continued to maintain, for another ten days, that the children were still likely to be alive. Their bodies were eventually found, at the spot where Chris had marked them on his map, by the perimeter fence.

The CD that Chris gave me was full of surprises too. The most interesting programme on it was ITV's Strange But True?, hosted by Michael Aspel, which was shown in October 1994, in it there was a reconstruction of Chris's warning to RAF Stanmore about a future terrorist attack. So certain had Chris been that the base was going to be an IRA target that he had got into his car, driven up to Stanmore and presented himself at the main entrance, to tell his incredible story to two bemused RAF security men. It must have taken considerable courage to do this on his own initiative.

In Chris's dream language terrorists are always portrayed as dogs and he had been having dreams about dogs in a graveyard, clocks ticking and photographic equipment, but more importantly by now he had discovered a vital new clue to making his dreams more effective. His spirit guide at the time, a dead soldier, had helped him to interpret where things were going to happen by hinting at post codes in the dreams. Chris was soon led to believe that the RAF base at Stanmore was the target for an IRA bomb. The log book at the base recorded his visit, when he was arrested and interviewed by duty sergeant Brian Earl, who also appeared in Strange But True? The airman was naturally sceptical, not to say suspicious. Who was this stranger and why was he making these outlandish claims? He wondered if Chris could give him any idea of timescales, but Chris explained that his information was never that precise. Chris was, however, able to give the details of his contact, a Chief Inspector in the Bedfordshire police. When the RAF was able to contact him he confirmed that Mr Robinson had been working with the police for some time. After three hours of questioning by RAF Officers Chris was released into the custody of the local Wealdstone police, who searched his car and then took him to the police station. There he was interrogated by a decidedly unfriendly Detective Inspector for another three hours, until a faxed message was received from Commander George Churchill-Coleman, who was head of the Anti-Terrorist Squad, it ordered Chris's release, and later on he was driven back to his car.

For a week after Chris's warning they doubled the guard at RAF Stanmore. Just over a month after his visit a bomb went off at the base, in an administration depot, which had been used to store photographic equipment. Chris's first reaction was shock and horror, but he was relieved when it was revealed that no-one had been injured and two people were taken in for questioning. All the main elements of Chris's dream were accurate – dogs had represented terrorists, clocks ticking represented bombs and photographic equipment had been in the building that was attacked. It was thought that the terrorists might have got through a back perimeter fence – which adjoins a churchyard, also featured in the dream... Dr Keith Hearne, a premonitions expert, was interviewed on the same TV programme, and in his opinion the psychic's premonitions were 'beyond coincidence.'

The second Robinson case to feature on Strange But True? concerned Dan Eldon, a twenty-two year old news photographer who had grown up in Africa. While he had been filming an attack by American helicopter gunships in Somalia the mob had turned on him and killed him. A few weeks earlier Chris had warned Cathy Eldon that her son was in grave danger. Chris had dreamt that a swarm of bees had surrounded Cathy's family (in Chris's dream language bees always translated as black people). Elements of the dream, as usual, had made little or no sense to the psychic. There were four dead photographers, a seaside place, F4 and 'rescue me Cathy' among the notes in the Dream Diary. It transpired that Dan Eldon had been killed at Mogadishu, which is a seaside town in Somalia, one of Dan's last pictures had been of female marines in bikinis with their rifles over their shoulders, four photographers did, indeed, die there and F4 related to a setting on Dan's camera. Dan continued to send messages to his mother from the grave, via Chris, showing where his negatives, many thought lost, could be found. The dead photographer wanted money raised and a charity to be set up in his name, all of which his mother, herself an ex-journalist, was able to do, partly from the proceeds of a book of Dan's photographs which she had successfully published.

The third story was about a recurring dream that Chris had about an airfield, aircraft looping the loop, explosions in the sky and two parachutes descending. This time he was able to witness the event first-hand. Chris told his friend, astrologer Penny Thornton, that he had heard about an air show that day and was going to test out his dream. It was the biggest military aircraft show in the country, the International Air Tattoo, held at Fairford, Gloucestershire. Chris rushed over and was in time to witness two Russian MIG fighter-jets in a mid-air collision that looked impossible for the pilots to survive. He was, however, able to reassure startled onlookers around him that it would be all right, as in his dream he had seen two parachutes, and so it turned out, both pilots safely ejected and were only slightly hurt.

The precognitive dreams all began for Chris Robinson in 1989, when he was 38 years old and living in extremely stressful circumstances. His video hire business had collapsed through no fault of his own; he was living in a caravan in a mobile homes park in Bedfordshire, and was involved in court cases about his ex-business. He was doing odd jobs to survive and had just become a father for the second time. One night he was awoken by what he thought was a child's crying, but then recognised the voice of his grandmother – but she had died several years ago. 'Christopher', she said clearly, 'someone is trying to steal your car.' Chris was sleeping at a friend's house, and his car was back at the caravan, five miles away. He replied, 'Nan, the car's five miles away. There's no buses, Lorraine hasn't got a car, and by the time I get a taxi or walk there it will be long gone. No it's lost.' 'Anyway', he added, wanting to get back to sleep, 'it's insured.' 'Don't worry, Christopher', his grandmother said, 'I'll do something about it.' The next afternoon as he reached his home a neighbour came out to see him and asked him if his car was all right, he was instantly reminded of the previous night's dream. 'Are you sure your car's all right?' she continued. 'Only something strange happened in the middle of the night. There was a blinding light, and a voice. My husband got up, because it frightened the life out of him, and he looked out of the window. There were these blokes trying to break into your car, so he chased them off. He spent the rest of the night sitting up, guarding it.' Chris thanked his neighbour, checked his car, which was undamaged, and then he went down to his local church, prayed to his grandmother and grandfather and thanked them for saving his car. He didn't know it then but this strange event was to be the beginning of his long psychic odyssey...

It was several months after this that, in his dreams, he met Robert, an ex-soldier who drowned after falling overboard on a ferry bound for Sweden. 'The soldier' was to feature regularly in Chris's dreams, becoming his main spirit guide to another world. Chris was subsequently to have other spirit guides who were not so anonymous... Unfortunately most of Chris's dreams are lost for that year, as he scribbled them down on bits of paper which then got lost. The next year, 1990, he meticulously noted down all his dreams in the Dream Diary, much of it is automatic writing, carried out while Chris is actually asleep!

One of those carefully documented dreams took place on Sunday, 13th May, out of eight or nine dreams that night one was startlingly clear, needing no interpretation of abstract symbols. Chris found himself standing by a railway line that he recognised as being local to his home.

It is part of the Metropolitan Line that runs from London out to Hertfordshire and into Buckinghamshire. This stretch goes between Chorleywood and Rickmansworth and there were four people with him, explaining that they were going to die here, but they seemed resigned to their fate and were not distressed. They added that it would be an accident which would occur during their work at this spot. This premonition was tragically validated just three days later. On Wednesday, 16th May, at 2 am, a team of five London Transport track engineers were taking part in routine track maintenance, a mile east of Chorleywood, when disaster struck. An 18-ton trailer wagon containing equipment and sleepers rolled over the gang of men as they worked on the line, the brake had slipped loose on the wagon, giving the men no time to clear the track before it was upon them. Only one man managed to scramble to safety – the other four were crushed to death by the rolling stock.

One branch of the law enforcement agencies that Chris has never had a problem with is HM Customs and Excise, who listened to him from the first time he contacted them, and have always acted upon what he has told them. An example of their joint success is illustrated by the following true case from Chris's files. He awoke one night sweating heavily and almost expecting to be in a prison cell, after a particularly realistic dream. He checked his Dream Diary and found that he had recorded it all, phrases about drugs and Air India, with a flight number, a day, and, best of all a description of a suitcase, with a drawing of it. He had dreamt that he was being awoken by a stewardess, and when he asked her what was going on she had told him it was time to disembark, they were at Heathrow. On the airport bus to the terminal he heard, quite clearly, the driver refer to the flight number, and then he told a nearby passenger that it was Thursday. When they got to the terminal Chris realised that he had no idea what case or holdall he was looking for on the carousel, so he stood and waited while the other passengers collected all their luggage, until there was only one case left. It was a new leather case with dark brown panels and a tan surround. He picked it up and walked towards the green channel. 'Excuse me sir', said a customs official, 'could I have a look at that?' 'But there's nothing in there,' Chris protested. 'I'll be the judge of that, sir,' said the official. The case was opened and inside was a top layer of clothing, badly packed. 'Well, well, what have we here, then?' muttered the Customs man. Underneath the clothes were a series of transparent plastic bags, filled with a brown powder – Chris just knew it was heroin. 'That's nothing to do with me,' he said, shaking his head as the Customs officer advanced on him...

So real had the dream been that Chris immediately phoned his Customs contact, Pete, (not his real name) who told him to fax the details then and there to his office. Some time later Chris found himself at Heathrow Airport, his contact had arranged for him to become a (temporary) baggage handler, complete with uniform. On the Thursday of the week that Chris was working at Heathrow Pete greeted Chris with a smile. 'You look pleased with yourself,' Chris said, knowing what Pete was about to say. 'You're the one that should look pleased, my son. It's Thursday, the flight number you wrote down is due in this afternoon, and although it's not an Air India plane, it is coming in from Bombay.' There were 800 cases on the flight and it took Customs a solid hour to narrow down the choice to nine cases, which were laid out in front of Chris, all of them in dark brown leather, and some had a tan trim. One, however, stood out, it was nearly new and Chris was sure that it was identical to the one he had carried through Customs in the dream. But there was a twist in store. 'That's the one in the dream', said Chris, 'but I reckon the drugs are in the one next to it.' He pointed to an old battered case that stood beside his 'dream case.' 'You sure?' Pete frowned. 'Why the change of mind?' 'I don't know – instinct, I suppose. But that's it, I'm sure.' The contents were emptied out; a false bottom was located and carefully lifted up to reveal several plastic bags filled with a brown powder, which was then tested. Pete was delighted, 'Jackpot, it's "smack", nice one Chris.' The courier stood trial and received a heavy prison sentence as a result of Chris's prophetic dream.

One of the most senior Establishment figures that Chris managed to persuade about his powers was Graham Bright, Parliamentary Private Secretary to John Major, which was just as well because Chris was having a series of dreams about an attack on Major, who was soon to become Prime Minister. The earliest warnings came in October '90 with clear intimations of a Cambridge post code (John Major was then MP for Huntingdon) and IRA involvement. By November a white Transit van was appearing as another dream feature. A dream on the 16th

November included Nigel Lawson, the Chancellor of the Exchequer, which led Chris to suspect that Downing Street, where the Chancellor has his home, at number 11, was the target. The November dreams also had references to a motorcycle and Chris deduced that the terrorists would make their escape by motorcycle. In early December Chris asked the spirits to show him where the next IRA attack on England would be. References to 'the PM' and to CB postcodes recurred again and again. 'Whole house shakes', from his notes, convinced Chris that there would be an attack on a house. The intensity, frequency and emotional impact of these messages were unlike anything he had experienced since the Stanmore incident, and they were driving Chris to the edge of madness as the months wore on. In January there were images of a white Transit van being resprayed by the IRA and snow (imminent danger) all around, as well as a drawing (from his notes) of a house, with the word 'MALL' written on the side of the house, as though along a road running in front of it. In February Chris dreamt about a roadblock, he asked the driver of a white van why the roadblock had been set up and was told that there had been a crash, leading to the roadblock. They had been driving too fast, killed themselves and the car was a write-off. Chris's interpretation of this was that there would be two terrorists (he had written '2 fast' in the diary) behind the attack. The vehicle from which the attack was launched would be written off and it would be a white Ford Transit van.

The whole affair reached a climax on Wednesday 6th February, with a dream about three rockets firing into space and the post code SW1. The intensity of this dream persuaded Chris that the attack was extremely close at hand – either later that same day or the following day at the latest. Chris woke up at 4am, unable to sleep any more. It was around 11 o'clock that same morning, and he was having a soak in the bath to try and relax, when the phone rang. He was disinclined to answer it, but a voice in his head said, 'Answer it Chrissy, it might be important,' and he had learned by now not to ignore any voices in his head! It was Graham Bright. He told Chris that there had been a mortar attack on Downing Street at 10 o'clock that morning. Three rockets had been fired from a white Transit van; the vehicle had then exploded in the Mall. Two men were seen making a getaway. One of the rockets had landed in the garden of No.11 Downing Street; the other two had hit No.10, but had caused little damage. It was later learned that the two men had made their escape on a motorcycle.

One of Chris's spirit guides was, and remains, particularly well known to the general public – Yvonne Fletcher, the woman police constable who was murdered by a cowardly sniper outside the Libyan Embassy in London in 1984. What is a spirit guide? There are a number of theories, chief among them being that they are spirits of the dead who return to guide the living in some way, often with information that would not be easily available to those who live in the material world. There is another school of thought that says they are entities that take on the form of the deceased, and a third theory maintains that they are the products of the human subconscious. Chris is firmly in the 'spirits of the dead' camp and I also subscribe to this point of view. Yvonne Fletcher was to feature often in Chris's dreams; she was involved in a number of foiled IRA attacks and in predictions of IRA hits, the Dream Diary recorded a conversation with her on the night of Wednesday June 5th 1991. 'Listen', she said urgently, 'there's going to be something about me in the paper tomorrow. I was with the Met until I died, and there's going to be a large cheque.' This meant absolutely nothing to Chris at the time as he didn't know Yvonne; all he knew was that he had been spoken to by a very insistent woman wearing a police uniform. The headline in the next day's Evening Standard took his breath away: YVONNE'S MURDER: GADAFFI ATONES Huge cash payment. Then he saw the head and shoulders photograph of the young policewoman, which was captioned: 'WPC Yvonne Fletcher: she fell dying into the arms of her fiancé Michael Liddle. People still leave flowers at the spot.' Chris instantly recognised the photograph of the WPC who had spoken to him in last night's dream. He had almost forgotten about Yvonne Fletcher, who had been in the headlines seven years ago. Chris read on and the story came back to him, WPC Fletcher had been policing an anti-Gaddaffi protest outside the Libyan Embassy when a shot rang out from a first-floor window, hitting the policewoman in the back and killing her. A Libyan official began to pepper the square with shot, injuring eleven protesters. The Embassy remained under siege for eleven days, Britain severed all diplomatic ties with Libya, but none of the Libyans was ever charged with the murder. The thirty officials claimed diplomatic immunity and were expelled without arrest. The large cheque referred to (blood money) was a six-figure offer of compensation that Colonel Gaddaffi made to a police charity.

Chris's theory about why WPC Fletcher had come back was that she wanted to carry on her work – she had been a conscientious officer who had loved her job. Later in June she came back to warn Chris about a bomb attack at the time of the full moon, 27/28th June, along the M4 corridor, somewhere near Heathrow. On the night before the attack Yvonne, looking pleased with herself, appeared in Chris's dreams to tell him that she had taken care of business. She was patting the breastpocket of her tunic. 'I've got the fuses, she said with a smile. 'I've got the fuses here in my top pocket. The bomb can't go off without them, can it?' They were standing on a piece of waste ground and Yvonne was laughing. 'The fuses on the timer, I've got them here. She patted her pocket again. 'No bomb's going to work if the timer isn't set, is it?' 'But how did you manage that? I mean, what happened?' Chris replied. The spirit giggled, 'I haven't been able to get through to you exactly where this bomb's going to be, right? But that's all right, because I've already told you that I'll see to it myself. So I followed these scum when they set out to plant it. They came round here' – she gestured at the waste ground around them – 'and left their van. The guy who was supposed to set the bomb brought it over here, and started to set the timer. Well I was going frantic, because I didn't know exactly what to do to stop him. He was about to set it, so I knew I had to do something fast: I did what any spirit would do. I crept up behind him and went "boo" in his ear'. She burst out laughing, 'I've never seen anyone run so fast in my life – or whatever you call this now. He was so shocked that he just dropped the fuse for the timer and legged it out of here, right over to that fence' – she pointed to a distant wire fence – 'and into the van. They shot off like the proverbial bat out of hell. And I just scooped up the fuses and put them in my pocket.' She smiled broadly, 'I'd like to see him explain that to the rest of the cell.'

The night of the full moon – 27-28th June was an anxious time for Chris, as he sought confirmation about the bomb. It wasn't till he switched on the midday news that he found the answer – the first news item concerned a bomb that had been found in the Beck Theatre, Hayes, less than a mile from the A4 as it runs past Heathrow. It had been found defused. Chris rushed to the phone to check with Sergeant Glen Clemence, one of his Bedfordshire police contacts. The Sergeant was able to tell him that the bomb had been found during a routine check, because the Band of the Blues and Royals – a military band, and likely target, were due to play at the venue. The bomb had been planted at the back of the theatre and had been primed to explode: it was a fully functional bomb in every way, yet it was perfectly safe. Chris asked why and Clemence laughed as he told him that the start button on the timer hadn't been pressed: a digital watch was being used and the stopwatch function used to time the bomb. A time had been set, but the button to start the countdown had been left unpressed. 'God knows why they forgot that,' Clemence said, 'unless...' 'Unless it was Yvonne,' Chris finished. Shortly afterwards, when he saw her in another dream, Chris asked Yvonne why she hadn't told him exactly where the bomb was planted. 'If I'd told you then there would have been people chasing around all over the place, looking for it. I wanted this one for myself. You see, Chrissy, the future was set: that bomb was going to be planted, and you couldn't stop that. What you could do was stop it going off. But I wanted to do that: I wanted to scare this monkey – and did he run when I whispered in his ear. Over the fence, into his van, and away. I don't know what he told the others, but he better have had a good reason why he didn't set the timer – after all, who'd believe that it was a ghost?'

One television programme, in particular, that Chris Robinson appeared in boosted his reputation as a genuine psychic enormously. It was This Morning with Richard and Judy, broadcast on the 28th October 1994 and it was ITV's daytime flagship show, hosted by Richard Madeley and his wife Judy Finnigan, which was broadcast live from a Liverpool studio. It remains one of the most popular daytime shows in Britain. Chris was appearing to promote the start of the Strange But True? series, which he was also featuring in later that same day. Chris had been set a challenge; he was to describe the contents of a sealed box, which would be opened during the live screening. Chris travelled up to Liverpool on the night before the programme, as arrangements had been made for him to stay the night at a hotel in the city. He only had the one night to dream about the contents of the box, usually he likes to have at least three nights of dreams before drawing firm conclusions: the frequency is an important establishing factor in divining the precognitive elements of his dreams. That night he dreamed about a phone box, inside the box was Raymond, whose nickname was Dolly, making a phone call. Another friend, now dead, Trevor Kempson, stood behind Raymond, waving a small doll at Chris. 'This is what it's all about Chrissy,' he said. The scene shifted, and Trevor was emptying a post-office sack onto the floor. Lots of boxes, wrapped up as Christmas presents, tumbled out of the sack.

When Chris was shown into the studio next morning he was confronted with a burly security guard, who had stood over the box all night. There was an excerpt from the Strange But True? film as Chris was led onto the set, accompanied by the security guard and the sealed box. Richard Madeley explained what the box was doing on a table and Judy asked Chris some questions about himself. She also asked the guard if the box had been in sight all night, he confirmed that it had. Chris then went on to describe the previous night's experiences. 'In my dream I saw a man in a phone box. This man was my friend Raymond, but we always call him "Dolly". So from that, I'd say that there was a dolly in the box. But where I've written down the word dolly as I sleep, I've put a box around it, and that usually means it's "not quite" what I've written.' Chris had the Dream Diary sheet in front of him to prove what he was saying. He continued, 'Later on in my dream, I was with a man who was emptying a post-office sack, and this sack was full of boxes. They were wrapped up like Christmas presents. Boxes again, and like presents… I'd say that there was a children's toy in the box, possibly a dolly.' 'Well we don't know what's in there, so let's open it up and see,' said Richard. When the box was opened, Richard pulled out a teddy bear – small, old and battered. 'I bet that is either called Trevor or Edward,' Chris added promptly. He was thinking of his friend Trevor Kempson, whose full name had been Trevor Edward Kempson. 'Hang on…' Richard paused as he received a message through his earpiece from the producer. 'Yes right…' He turned to the camera, 'The teddy bear belongs to our producer, Helen Williams, and it is called Edward.' The interviewer's amazement was plain to see, 'That was remarkable.' 'Well I did say a dolly, or like it. I didn't actually say a bear', Chris replied. 'But Helen's just told me that her parents used to run a post office when she was a child, which could be where the post-office sack comes in. What's more she was born on Christmas Day, which would account for the presents in the sack. And you knew what the name could be.' Chris had scored three hits – a child's toy, named Trevor or Edward and the link between Christmas, a post office and the contents of the box. Once again, this was a long way beyond mere coincidence...

Chris was later contacted by Marie McCourt, who had seen him on the show and who needed his help to lay to rest the disappearance of her daughter Helen. It was known that Helen had been murdered, a man had stood trial, was convicted and sentenced, but he had never revealed where her corpse had been hidden. Now, six years after the conviction, it was time to find Helen's remains and to give her a proper burial, if Chris could locate her. What Marie didn't know was that two days before he travelled up to Liverpool Chris had been visited in his dreams by a young girl, who told him that she was lost, and had been murdered, she wanted her mother to know that she was well, and to help her mother to find where she was buried. When Marie McCourt wrote to Chris he was able to ring her and tell her about the girl he had seen, with a description and some details she had given to him. Marie, tearfully, admitted that this was indeed her daughter; things he had relayed could only have come from Helen. Marie's idea was for Chris to visit her in Merseyside, to sleep in Helen's bed and to see what his dreams revealed, to which he agreed.

In the meanwhile the psychic detective was having dreams about prison, he was in a prison cell, but the man with him seemed unaware of his presence as he sat looking at the wall in silence, every muscle in his body tensed. At ten o'clock at night the lights went out, the man rose up; faced the door of his cell and in deafeningly loud tones quoted verses from the Bible in a declamatory tone. It was a sing-song voice, almost a meaningless chant of Old Testament verses, droning on and on about who begat who, going on and on, into the night.

Helen reappeared one night, to take Chris to some waste ground, up a steep incline. At the top was a view of five ponds. 'I'm in there,' said the spirit, but Chris was unable to establish which one, however recurring letters in his Dream Diary revealed a postcode and Chris discovered that it covered a small area around Billinge, a village near St Helens in

Lancashire. Soon after this Chris received an offer of help from a police officer named McKay, who was familiar with the McCourt case. When Chris relayed the information he'd gleaned about Billinge and the five ponds McKay was able to identify the area as Moss Bank – a couple of miles from Billinge village. Chris also revealed his dream about the prison cell and the Old Testament ranting and raving, which had been bothering him, he'd wondered if it had anything to do with Helen McCourt's murder. McKay was momentarily stunned into silence, eventually he was able to respond. 'How the hell could you possibly know that? 'The man we've got banged up for the murder – you know what he does? Every night, at lights out, he gets up in his cell and starts shouting verses from the Bible. It drives the other prisoners mad, and most of the staff. He shouts so loud that they can't ignore him, and he does it every single night, without fail. He has done so ever since he was convicted.'

About two weeks later Chris travelled up to Liverpool, stayed at the McCourt home and dreamed about Helen but gained no new information, only a repeat showing of the five ponds. He met McKay, the two men visited Moss Bank and Chris recognised the site from his dreams. Chris's theory was that Helen couldn't tell him which of the five ponds her body lay in, the terror surrounding her abduction and murder would have made her confused. It was not a priority for the police to drag all five ponds – the killer was already in custody and dragging five ponds would require an awful lot of manpower…

There are two big questions that may never be answered concerning Chris Robinson. How do the messages come to Chris in his dreams? Why should Chris and nobody else have this gift, at this level of intensity? Chris believes in life after death and he believes that his messages come from the spirit world, a world beyond death. Having got to know Chris I can say this, he is a good man, with a conscience, he is not motivated by money and he believes in God. He has tried his best to make sure that his gifts are used to serve his fellow man and this has been at some considerable personal cost. He is not precious about his genuine abilities, indeed he stresses the fact that we all dream, and we are all capable of recording our dreams and developing our ability to analyse our dreams for the messages that they may contain. The human race remains woefully ignorant about how and why we all dream; more research needs to be devoted to this fascinating subject.

There have been some suggestions made as to why the gift of precognitive dreaming should have been granted to Chris Robinson, including the possibility that he is somehow more developed mentally in that area of the brain that controls our dream state – we still have much to learn about the workings of the human brain. There are a handful of people like Chris throughout the world, I believe that they should be brought together and studied by scientists, so that any similarities between them may be identified and their abilities encouraged and developed where possible. If one more life can be saved by their warnings then it will be time well spent.

Another suggestion about Chris's gift, made by his parapsychologist friend, Dr Keith Hearne, is that the prolonged periods of anaesthesia, endured as a child during open-heart surgery, may have affected him, or the many near-death experiences over the years, including the four month stay in hospital for the surgery when he was nine and the serious injuries he sustained in a motor scooter accident when he was sixteen. Many people who have been studied by Dr Hearn, that have psychic experiences, also have undergone prolonged periods of anaesthesia.

It has been a painful, lonely journey through uncharted waters for Chris Robinson, a test of character that he has passed through. He has worked tirelessly to crack his own personal dream code and to get his dreams acted upon. I know this much is true, Chris will never give up on helping people, on developing his ability and continuing to look for answers to the many questions raised about his rare and strange gift of prophecy...

Dunstable Visions

Inspired is a New Age shop at 26 Middle Row, right at the heart of Dunstable, which has a story to tell. Over the years it has seen various changes in use, before it was called Inspired it was Mystique (also a New Age shop) and before that it was World of Enchantment, in the

1950s it was a fancy pottery and handbag shop and in the 1940s it was a butchers shop. Directly opposite stands Albion Buildings, formerly a hat factory, and stocks used to stand on the street outside number 26 Middle Row. The present partners – Caroline and Sam, have been here since 2000 and have had a number of strange experiences. I met Jean Peck, Sam's mum, and a psychic medium called Graham Stenhouse, who related some of the occurrences when I called in during my researches on Thursday 26th August '04. Galaxy Television had heard the ghostly stories surrounding Inspired and had decided to make a programme about the vigil held here on a cold November night some two years previously. Unfortunately I hadn't seen this programme, but what Jean had to tell me gripped my interest. Jean has heard her name called when there was nobody nearby, she has also witnessed the electric lights flickering while some bamboo wind chimes were violently smacked together, on a windless day when the doors were both shut. A statue has been seen to fly straight through the air and to break on the floor. Graham told me that the place used to be a foster home and he has seen the ghosts of children waving from the upstairs windows and there is a cold spot in the upstairs room. He has also sensed that a man hung himself on this site, in the mid 1800s, after being unjustly accused of a crime he did not commit. He has seen two women downstairs who haunt the corner of the room where they sit together. He has also seen 'a big guy with big hands' who carried out an abortion on his niece, which involved some high-ranking church person, down in the claustrophobic cellar, which I have explored; it is accessed via an extremely narrow set of steps.

I returned to Inspired seven weeks later, to talk with both of the shop's owners, Caroline and Sam and their friend Denise Newman, who is a 'spirit communicator'. She has had many communications from the spirits that inhabit the Inspired premises. I also met Sam's dad Ted, he told me that he hadn't believed the stories until he had seen a troll, a heavy plaster cast model, lift up from its stand near the display window at the front of the shop, and drop to the floor. This was on a summer's day, about eighteen months ago. He had also witnessed the wind chimes suddenly ringing one afternoon when all the doors were closed in the shop. It's always good to get the testimony of a former sceptic! Samantha has felt something 'brushing against my legs, my hair has been stroked, and last Christmas, when I was upstairs my bottom was poked.' She also told me that candles have been found all mixed up and she had come into the shop to find things strewn across the floor. The atmosphere can change abruptly here and there have been odd occasions when Sam has felt so uncomfortable in the shop on her own that she has called in Caroline. There is also an unaccountable 'disgusting smell' which several people have encountered from time to time, and Sam described it as 'like sulphur.' Both partners agreed that if they ever bad-mouthed anyone during conversations the lights would flicker, as thought the spirit people disapproved. The girls also told me that if things were really quiet they would ask their invisible friends to please look after them, invariably business would soon pick up and the money would start flowing in again.

I was interested to learn about the séance that had been held on Friday evening a couple of weeks before my visit. Caroline explained that her back had been cold, though everyone else felt warm, for the entire day of the séance. During the séance there were four people, Denise, Caroline, Marie and Jean and everyone had complained that their legs, from the knees downwards, were freezing cold. The glass they were using moved slowly at first, but as Denise meditated it got quicker until it was, according to Caroline, 'flying around the table, very fast.' The group picked up Sara, one of the children who haunt the shop, an eight year old who claims that Caroline was her mother in a former life. Denise told Caroline that the little spirit wanted to talk to her, so Caroline did a lot of the questioning. Another séance is planned for the future as it became clear that Sara had been delighted to be able to communicate with the living.

Even more astounding revelations were in store. On the Friday before the séance was held both Denise and Caroline had sat upstairs together and when they came down the scene on one of the CCTV monitors had changed. It was showing shelves and storage boxes instead of the usual interior of the shop. As Denise and Caroline watched two figures materialised. There was a man on his back with his knees drawn up and his mouth agape as though he might be drunk or injured. A little girl appeared, bending over the man but looking directly at the camera. Caroline was amazed by the image, 'It was perfectly clear, she had dark eyes and dark hair, parted in the middle and worn down to her shoulders, she had on an old-fashioned pinafore dress and would have been about eight or nine years old, I didn't feel scared by it.' The girls noticed the images at ten o'clock and they were still showing on the screen at eleven o'clock when they locked up the place, sadly the cameras don't record anything, and to capture such an image would have been invaluable.

I asked Denise for her thoughts on Inspired and she corroborated Caroline's story about the CCTV images. 'There are lots of children here, I smelt smoke at one time, and it turned out that there had been a fire long ago, which affected the properties on either side of this building, but this place didn't actually catch fire, so there would have been strong smells of smoke then. I often feel the children tugging at you or poking you. I have sensed the presence of a priest, carrying out an exorcism, a little boy with callipers on his legs and an old lady called Alice. When I was upstairs with Caroline white stuff appeared on my trousers from nowhere, it looked like flour, and the children may have played with flour. We have heard the sounds of pebbles being thrown into a corner up there.' Denise went on to say that she has also contacted the spirit of Sara through automatic writing at her home. The messages are always drawn in 'bubbles', like cartoon speech bubbles, with words like 'shop' 'love Caroline' and 'Grace', who was Sara's friend, written inside them. Only the other day Denise was in the shop when she felt her coat yanked. The spirits of Sara and her friends seem to be happy to remain close to Caroline in particular, so Denise has no plans to move them on.

One of the oddest stories associated with Inspired was the sensing of two black slaves, by a medium during his visit to the shop, he had journeyed a long way to investigate Dunstable. Nobody could understand what black slaves might possibly have had to do with Dunstable. Well-known local historian Vivienne Evans decided to do some research, using the history book Dunstapleogia, and she made a surprising discovery. In 1845 William Wells Brown had held a Bible Society meeting in the Temperance Hall, West Street, about slavery. The guests of honour were two escaped American black slaves! It is, therefore perfectly possible that they visited Middle Row... The same medium picked up John Dee's presence in West Street, he knew of Dee because he had encountered his spirit before, on the Welsh borders. Once again historical research backed up the possibility. Dee had been a friend of Queen Elizabeth I, the Queen was known to have passed through Dunstable on many occasions and she might well have been in the company of her friend John Dee.

Earlier on that Thursday in late August I met Mandy Cummins, a barmaid at the White Horse/Finbarrs, an old Dunstable inn, and her partner, Michael Whitehill, who had gained an unusual 'attachment', this is the name that parapsychologists give to a ghost that follows you home. Michael had been drinking in the pub about a week prior to our meeting and he was touched on the shoulder, but when he turned around he was alone. He got home, but the entity followed and was still residing with Mandy and Michael when we met . In their kitchen both of them had heard a whistling noise and searched for the source of it, they soon realised that their invisible friend was blowing the dog whistle which had been hanging up on a hook. Later Michael was prodded while in bed and the spirit took a liking to a knob on their stove, it disappeared and has not been found again. Their nameless ghost from the White Horse is a young girl, according to Michael she is fifteen and a half, wears a black dress with white apron and peaked cloth cap tied at the back and earned her living as a dray girl. She has walked through him and touched him, she makes his hair stand on end and plays with candles, elongating the flames and making them smoke, but neither Mandy nor Michael is at all bothered by their unusual 'houseguest'…

Whilst on the subject of pub ghosts, The Norman King deserves a mention as it also has a reputation for being haunted. It is only relatively recently that the building has been used as a pub – during the past 42 years, a mere blink in time for a place with a history that dates back to 1106. Over the centuries it has had numerous uses; these include stables, a library and a museum.

I spoke with the landlords, Jed and Mel McGarry, but the main source for the haunted history was Pat, who has worked here for the past 20years. She showed me the large fireplace, built of brick and stone, where a big vase of flowers had fallen from the high ornamental top, about the height of a man, onto a solid wooden floor, without breaking – it simply bounced. Pat's daughter, Elaine, who is six, told her mother that she could see a little boy, 'making things wobble.' The Guinness pump has been known to turn itself on and off too. Pat went on to explain that it was mainly the restaurant end of the pub, and the kitchen area in particular, that was haunted. Dogs have been known to refuse to enter the restaurant, their hackles rise if they are led near to it. She took me inside the kitchen where a straight corridor leads to the storage area. The door we had just come through often opens of its own accord in the evenings, and it is quite a heavy door. Near the cooking range it frequently seems as though an invisible presence passes by, along the corridor, reflections of this 'presence' are even glimpsed in the fronts of the microwaves. The fridge has been turned off, yet to turn it back on again meant pulling out heavy units to get to the plug. Sometimes the table creaks as if something heavy has just been put on it and when there is no-one in the storeroom objects fall off the shelves. Interestingly, Thursday evenings are a time when things will regularly go wrong in the pub. Pat remembered that one of the waitresses had done some research on the site some years back and found that the building was once surrounded by a moat and that a boy had drowned in this moat.

(I am indebted to F A Fowler's historical notes – The Priory Church of Saint Peter, Dunstable – A Brief History, for the following facts about Dunstable Priory). King Henry I founded Dunstable town around 1111and built his 'palace' of Kingsburie there. In 1131 the Priory of St. Peter was established, under the leadership of Prior Bernard, who was associated with the introduction of the Augustinian Order into England It was sited opposite the King's home, and he granted the Priory lordship over the whole town, which was to lead to enmity between townsmen and Prior in times to come. Soon after the endowment was made the building of the church and convent began, but it was to be another 70 years before work was completed! Disaster was not long in coming to the attractive Norman church, in 1220 the roof of the presbytery fell and at the end of that same year a terrible countrywide storm felled the two western towers, one of which smashed in the western front, which had to be rebuilt, but without towers.

During the Middle Ages many alterations were made to Dunstable Priory, between 1228 and 1289 works included the building of Our Lady's

Chapel, the construction of an inner gate within the court, ten tons of lead were added to the refectory roof, a great stable was built, five large bells were made and two pinnacles were added to the front of the church. Dunstable continued to grow in importance, many Tournaments were held here and King Edward I, King Edward II and King Edward III all attended, with many of the nobility, straining the hospitality of the canons quite considerably.   King Edward I brought the body of his beloved Queen to the church, where it rested overnight.   By the crossroads, near where Eleanor's body had rested, a cross was erected to her memory.   On the journey down to London from Lincoln, where the Queen died, the distraught King had a series of crosses erected, marking the places where the cortege had rested.   The most famous of these were to create place names - Charing Cross and Waltham Cross, but only three of these beautiful crosses still remain.   The Dunstable cross, however, was to be destroyed by Fairfax's soldiers during the Civil War (1642-1649).

Prior Thomas Marshall (1351-1414) continued the rebuilding of the Priory Church, but, as part of the great peasants' revolt of 1381, Dunstable burgesses, led by innkeeper Thomas Hobbes forced a charter from the Prior, who gave in to the mob. He refused to revenge himself, as the local nobility did, when the revolt had been crushed.

The last Prior of Dunstable was Gervase Markham, elected in 1525. Archbishop Cranmer visited Dunstable Priory in 1533, with the Bishops of Winchester, London, Bath and Lincoln, to judge their royal master's (Henry VIII) cause in the matter of the divorce of Queen Catherine of Aragon – the end of the monasteries was fast approaching and just twelve days in May 1533 were to shake the English world... Dunstable had been chosen for its proximity to Ampthill, where Catherine was staying. The Court was opened in the Lady Chapel of the Priory Church on 10th May. Catherine was cited but did not appear and on May 23rd 1533 Cranmer declared the marriage null and void, the writing of the document was affixed to the church door. The Dissolution of smaller monasteries was to follow, with the Act of Parliament in 1536, and pressure was brought to bear on the larger houses until, by the spring of 1540, not one English monastery remained.   With no stain on its character the deed of surrender at Dunstable Priory was signed on 31st

December, 1539, and the monastery dissolved in January 1540. Both the prior and his twelve cannons were granted pensions and dispensations given to them to serve as secular priests.

After some few years the great church and monastic buildings were plundered for all that was valuable and left a ruin, to become a quarry for the whole neighbourhood. A wall was built up in the church from the rood screen and the nave and its aisles sealed off for the use of the parish. During Edward VI's reign the reformers spoiled the church still further, but in Mary's reign some order was restored, as the remains of the screenwork of that period testifies.

The next hundred years (1550-1650) brought religious upheavals with Reformation and Extremism in South Bedfordshire. When Cardinal Pole visited in 1556 there was no permanent rector or curate. By the end of the reign of Elizabeth I the Puritans were extremely active in parts of the county. Edward Alport, rector at Dunstable, suffered at the hands of the Puritans and Anabaptists, who, in 1616 'baptised' a sheep in the font and then placed the animal in the pulpit in mockery of both baptism and the preacher. They refused to pay the rector's tithes, cut down his corn and spoiled his crops. After he complained they beat him so badly he almost died, but went unpunished. Alport left in 1623 and extremist Zachary Symms replaced him, but he resigned in 1634 and set sail, with the rector of Odell, for Massachussets in pursuit of greater religious freedom. Next came William Pedder, a Royalist, appointed rector in 1634, who encountered much opposition and was turned out in 1642 then Parliament appointed a group of sixteen 'lecturers' to preach at the church on Sundays and weekdays. The town was raided one Sunday in June 1644, by Royalist soldiers, who forced the doors and fired their pistols, wounding several of the congregation. Ten years later Oliver Cromwell was petitioned by some parishioners for a minister and in 1656 Cromwell's men selected Isaac Bringhurst, the first rector the parish had had in fourteen years.

The Restoration of the Church of England in 1660 saw Bringhurst leave and a new rector, called Lyster installed, but in the later 17th and 18th century not much is known about the rectors of Dunstable. Some restoration was carried out to the church around 1720, but by 1845 it had become ruinous and in 1852 a drastic scheme of repair and rebuilding was carried out. There was further internal restoration in 1890-1891 and the west front was restored in 1903. Sir Albert Richardson directed repairs to the south-west buttress in 1930. Between1962 and 1989 a Dunstable family paid for many improvements including restoration and embellishment of the east end and sets of windows in five different locations were provided too.

Post war growth saw the establishment of St Augustine's Church in 1959 to serve the south of the town and in 1968 St Fremund's Church was built in north Dunstable.

On Saturday, 10th May 2003, I assembled a team from Anglia Paranormal Investigation Society for a night investigation, at the Priory House in High Street South, Dunstable. We were fortunate that the property developer, Oliver, and his wife Natalie, had invited APIS to carry out a vigil, as there had been some unexplained occurrences in the building, leading Oliver and Nat to suspect that the place might be haunted. We were not going to be disappointed.

The sense of history in the old building was overpowering, one could easily imagine King Henry VIII himself, striding through the hall and up the stairs. The empty building was quite dark, so we needed our torches once the lights were turned off. The other thing that struck you was the size of the place; it had about twenty-five rooms. There were eleven of us in the team, and we set up our equipment at 10.30pm and organised ourselves into parties of two, while I remained mobile, helping out where necessary. Those in the team who were new to ghost-hunting later reported that they hadn't known what to expect, they were both nervous and excited. My paramount concern for the team is always that everybody should be safe, so we had an evacuation procedure in place, we carried a first aid kit, and I had instructed the less experienced investigators in how to conduct a safe and successful vigil. I already knew the paranormal 'hotspots' where hauntings were most likely to happen because my friend and fellow APIS member, Keith Paull, had previously remotely dowsed the site, using a pendulum and a map. His 'code' for the type of energy lines which ghosts use is always green, and I carried a plan with several green lines and spots on it. I don't tell the team where these 'power lines' are situated, I also don't tell them about any previous sightings or other reports of paranormal activity. This would be likely to prejudice the investigation.

We had three 'watches', with breaks for refreshments – mainly coffee – and stayed in the building till 3.30am. As is often the way, when we were all relaxing during a break, after the first watch, things began to get really interesting. Donna, APIS's Secretary, was passing a small room on her way to the kitchen when she was startled to hear the sounds of a heavy object being dragged along the floor, so she came and got me and pretty soon the little 'office' was full of APIS investigators. Donna sensed a man in his late 30s or early 40s, who had died from a head trauma. Two of our group, Simon and Julie, also psychically gifted, picked up several entities and made contact with an unhappy married couple who had once lived in this place, the man kept handling his fob watch, leading us to suppose that he was from the Victorian era.

On the second watch, about 1 am, I was summoned, via the walkie-talkie, to the top floor, where my brother Paul and his team-mate, Simon, had seen a window shutter moving of its own accord, first one way then the other, on a perfectly windless night. We closely examined the shutter but could find no rational explanation for its movement.

The final watch had another surprise in store for us. Once again I was contacted on my transceiver, this time to the vaults, where Ollie and Tina had their camcorder set up. They were both complaining about the intense cold, a temperature drop which both of them had suddenly felt, although I, having just come downstairs, didn't notice it so much, and so I checked the temperature/humidity gauge, and it was going haywire! Such a thing had never happened before (and hasn't happened since) usually the temperature varies by a few degrees, gradually throughout the night. The humidity, similarly, will alter by a few per cent over the hours. This however, didn't make sense, as I looked on both the temperature and the humidity were changing too quickly to record (we usually take temperature and humidity readings every quarter of an hour) they were zooming up and down the scale crazily. The next thing that happened was the camcorder battery drained right down and it switched off – despite having a new battery fitted. The pair of investigators looked a bit spooked; something was going on in those vaults. It was only later that I was able to tell them that one of Keith's 'haunting power lines' ran straight through the vaults, right where they had been sitting... Another haunting line had been detected in the attic, running diagonally across the room and I was a little disappointed that it had been a quiet night up there, with little to report. Or had it? Some of the team had sensed the presence of children, and Donna picked up the name, which she thought was Elspeth, of a young girl of about 15 or 16 years of age, on the back 'service' stairs and the walkie-talkie reception had been impossible in a few areas of the attic, as it was affected by intense bursts of static. APIS member   Paul Watts had taken some photographs up there in the dark, and when they were developed a shadow could be seen on the back stairs leading up to the attic, it is a small silhouette, and it could just be that of a little child...

The Priory House has now been redeveloped and sold to South Bedfordshire District Council. I am hopeful about carrying out another vigil in this fascinating old building, which has played such an important role not only in Dunstable's, but also in England's history.   It is looking quite positive as I have been in discussions with the Council, who seem most encouraging, and it is probable that APIS will return for a follow-up investigation at Priory House – and who knows what that may yet reveal?

Chapter 2

Ampthill Apparitions

Ampthill is an attractive Georgian market town, on the Greensand Ridge, (Bedfordshire's premier forty mile walking route) which overlooks the Vale of Bedford. Lord Upper Ossory, of Ampthill Park House, was responsible for reorganizing the layout of the Market Square, and building both Ossary Cottages and Catherine's Cross. His nephew, Lord Holland, gave the town its splendid avenue of lime trees, the Alameda. This small town's main historical fame is its association with Catherine of Aragon; indeed The Queen's Head pub is one of the town's tributes to this wronged lady. In the fifteenth century King Henry VII sought to secure his position with dynastic marriages. Catherine of Aragon was the Spanish Princess chosen for his eldest son, Arthur. By the time Henry VII died, in 1509, Arthur was dead too. Henry VIII was the new King, and his decision to marry Catherine was partly influenced by the fact that he was unwilling to repay her dowry, which had been made earlier to Arthur. The marriage was, however, a happy one, marred only by Catherine's inability to produce a male heir. Of the five children she bore Henry, Mary alone survived. Henry began divorce proceedings in 1527 and eventually obtained an annulment in 1533. Catherine was exiled from court and went to live in the Manor House of Ampthill Castle, which was built in 1420 by Sir John Cornwall. The building was financed from the spoils of war with France, including the battle of Agincourt. Records of the castle speak of 'foure or five faire Towers of stone in the Inner Warde'. King Henry VIII was a frequent visitor. With the death of the King in 1547 the castle fell into ruin, and the good people of Ampthill helped themselves to this ready supply of building materials.

One act by Catherine of Aragon, fondly remembered by the citizens of Ampthill, was when she burned her lace, so that work would be available for the townswomen in replacing it. She died at Kimbolton Castle on the 7th of June 1536. Henry did not attend her funeral, held at Peterborough Cathedral. A black growth had been found on her heart but rumours persisted that she had been poisoned. Of Ampthill Castle, where Catherine learned of her annulment, all that remains is a stone cross, Catherine's Cross, erected in her memory in 1737. On it is the following inscription:

In days of old here Ampthill Towers were seen
The mournful refuge of an injured queen
Here flowed her pure but unavailing tears
Here blinded zeal sustain'd her sinking years
Yet freedom hence her radiant banner wav'd
And love aveng'd a realm by priests enslav'd
From Catherine's wrongs a nation's bliss was spread
And Luther's light from Henry's lawless bed.

Strangely Catherine is not reputed to haunt the town that gave her succour. Instead her ghost has been reported many times at Ludlow Manor in Shropshire. This was a place of happier memories, when she was married to the young Prince Arthur. Yet Catherine remained faithful to Henry, whom she had married when he was just eighteen, to her dying day.

The Great Park, an estate of 300 acres, where Ampthill Castle used to stand does have a ghost. It is definitely not poor Catherine of Aragon though. In 1965 an ex-policeman, on holiday in the area, was strolling through the park with his family. He reported seeing 'the vague shape of a man on a horse who suddenly appeared from where the castle used to be and vanished near a small brook.' Stories of a ghostly rider seem to have been connected with the Great Park for many years before this particularly well documented sighting.

In 1979 the author Kit Williams buried a jewel-encrusted golden hare and in his book Masquerade the writer concealed clues to the whereabouts of this treasure within a series of paintings. The book was responsible for sending some two million treasure hunters, from every corner of the globe, in search of the golden hare. Finally, in February 1982 the winner unearthed the prize, in the shadow of Catherine's Cross, Ampthill Park!

Houghton House, in Ampthill, stands on land originally known as Conquestbury or Houghtonbury. The manor of Conquest was held from the 13th to the 18th century by the Conquest family, and the manor house then stood in a secluded hollow near Bury Farm. Houghton, now a ruin administered as an ancient monument by English Heritage, was built on a hill to the south of Ampthill, in what had come to be known as Houghton Park. A three-storied stately home, it was constructed circa 1615-21, on crown land, for the Countess of Pembroke (sister to Sir Philip Sidney, the Elizabethan courtier and poet). The Countess did not enjoy it for long, as she died in 1624. The property passed to the Bruce family (Earls Elgin and Ailesbury), who took the wrong side in the Civil War and ended up in The Tower of London. In 1738 the Duke of Bedford bought the house and estate. By 1794 he was arranging its dismantling; much of it was recycled, for instance the great staircase ended up in Bedford's Swan Hotel. In the 1930s local subscriptions were raised to save the building for the nation, and today it is in the care of English Heritage. The ruins of Houghton House, in Ampthill Park, are open to the public, though the main driveway up to the house is now a ploughed field.

Some years ago an older resident of the town vividly recalled her ghostly experience of 1915. She was returning home, after war work at a factory in Luton (her parents had a small farm nearby); it was late at night when she heard the loud noise of many horses approaching. As the lane was quite narrow she didn't want to be injured, so threw herself into the ditch. There followed a tremendous thundering of hooves and jingling of harnesses, as though some great coach was sweeping up the driveway to Houghton House, but the lady was startled to see nothing visible to account for the tumult that had caused her to fear for her safety. Houghton House still continues to attract visitors, despite its ruinous state. Sam Waddington was one of those people captivated by the old site's atmosphere, along with several of her friends. She took some photographs during a visit a few years ago. Strange floating orbs appeared on the pictures, for which Sam could find no explanations;

they certainly weren't there when the photographs were taken. The phenomenon of these strange orbs is one that holds an interest for me. They look three-dimensional, rather like the product of a child's toy bubble-blowing pipe and are invisible to the naked eye. It is only when the film is developed that they become clearly visible; they are believed to represent a person's spirit – as an orb of light, or the beginning of a full-blown manifestation, a sort of 'embryonic ghost'.

I visited Houghton House, in the company of my friends, Julie Watts and Sam and Barry Waddington, on Thursday the 31st October 2002. Halloween seemed an appropriate time and sure enough it proved to be most productive for Julie, who is a practising medium. It was my first visit, but Julie had been here quite a few times before. She soon began picking up impressions; the first was of a small man, who had been a servant at the Hall, the second, in another corner of the ruins, was a stronger presence, tall and more imposing; she picked up the name Sam, and felt sure that this person had been in charge of the other servants, and although he seemed shy he was a benevolent presence. The third spirit was female, wearing a white gown, and appeared to be on a floor level that had once been above us; you are at ground level in Houghton House as all the staircases are long gone. She had been the mistress of the house, was most happy here, and Julie picked up the name Vanessa.   The fourth spirit was definitely not friendly; this unnamed spirit wore a wig, like a judge's, and was a person of rank, and he was hostile to our presence. Julie felt strongly that if we had stayed for an overnight vigil he would have made his presence felt physically, by touching us and making footsteps to frighten us as he would have wanted to cause mischief. Sam Waddington too, later picked up the same presence as a dark outline, brooding and keeping us under observation. It was after midnight and Barry Waddington and I had taken a few flashlight photographs. The girls thought it was time to go. We were intruders as far as the unpleasant male spirit was concerned and he wanted us gone, so we duly complied. It certainly is a romantic ruin, with good views of the surrounding countryside, well worth visiting, and I feel that the people who lived here were most happy to do so.

In August 1987 The Ampthill & Flitwick Times carried an article about a haunted house in Chandos Road, Ampthill, the home of Peter and Irene

Francis. Peter Francis was mayor of Ampthill at the time. The couple reported the presence of a mischievous ghost with an annoying habit of hiding things. Cutlery, jam, spectacles and jumpers went missing. A heavy sliding door moved of its own accord and bread and bowls fell off shelves. These odd occurrences had been going on since the Francis family moved in, back in 1968. The children came up with a name for the resident spirit – Mrs. Farr, and it was a family joke - until they discovered that a Miss Elizah Parr had once been a housekeeper at their Chandos Road home. Further research revealed that Elizah, known as Totty, worked in the town at the turn of the century, first at the White Hart as a chambermaid and later in Chandos Road. Rumour has it that Totty was caught up in a scandal at the White Hart – there was talk of a child. Totty had her own special chair in the Francis home, and reserved most of her mischief for the kitchen – the place she knew best. She liked to take pots of strawberry jam which later came back to the table. Totty was a ghost never heard, but she was seen, twice. The first time was by the couple's younger daughter who saw a woman standing at a window. The second sighting was by the mayor himself. 'I saw something at the end of the upstairs corridor – it appeared to be a figure in a long skirt'. Elizah Parr was buried in August of 1932.

The White Hart Hotel, where Totty Parr worked some seventy years previously, reputedly had a haunted room – number 13 appropriately enough. It was reached by climbing two flights of wooden stairs and walking down an eerie attic corridor, where the ancient roof beams were clearly visible. The White Hart is certainly old – parts of it dating from the 15th century. In 1997 Bedfordshire Times reporters spent a night in room 13. They immediately noticed that the door at the end of the corridor was held open with an ironing board, to indulge the ghost! Inside, the long thin room was bare with wooden floorboards and a tiny, curtain-less window, as the room was normally unoccupied. Although nothing untoward happened during their brief tenancy, they did have an unexplained experience in the back bar. At ten o'clock that evening, the window swung wide open, yet the wind was blowing in the opposite direction. The newsmen were convinced that no human hand could have opened it. One reporter was standing beside it at the time, and on the other side was a twenty foot drop to the courtyard below. Frank, a regular who was interviewed by the reporters, claimed two previous encounters with ghosts at the hotel. His wife had worked there as a cleaner. One day as she was cleaning, Frank came up and tried to close the door. 'My wife told me not to as 'she' doesn't like it. I put the door back, but to teach me a lesson it swung back and hit me on the head!' On another occasion Frank had been in the front bar when a man in a top hat walked in, Frank told reporters, 'Three or four of us saw him for a split second; he was dressed like an old coachman with a cape,' then the mysterious stranger vanished.

The landlord of The White Hart at the time, Shea Keegan, was in no doubt. He told the newsmen, 'There's definitely more than one ghost here.' He also said that glasses used to move around the bar by themselves in the middle of the night. It seems likely that 'Totty' haunted The White Hart as well as the house in Chandos Road.

Just seven months after the White Hart Hotel investigation, an even stranger location hit the headlines around Bedfordshire. Bedfordshire on Sunday reported on some haunted hairdressers! 1710 Hair Design and Beauty in Church Street, Ampthill was reputedly haunted by at least two ghosts, a badly frightened customer reported seeing the image of a hanged man in an upstairs room. Michelle Tollett, the 24-year old owner called in exorcists, which seemed to have the desired effect of ending the manifestations. There was, however, still the ghost of a young boy that was regularly seen upstairs by both staff and customers. Michelle commented, 'It is not very clear but there is definitely something up there and it can be quite daunting when you are on your own.'

The immediate area in and around Ampthill is highly likely to continue to attract ghost hunters for a long time to come. One day, perhaps, towns like York and Edinburgh may be challenged for their share of ghost-walk visitors by their much smaller (population 7,000 plus) but no less interesting southern country cousins in Mid-Bedfordshire.

Chapter 3

Spectres of the Skies

Thurleigh was once home to the 306th Bombardment Group of the United States American Air Force during World War Two. The 306th had the distinction of having the longest tenure of a UK base by any American combat unit: arriving at Thurleigh in September 1942 and departing in December 1945, after flying 342 missions in its B17 bombers. Sadly, the 306th had another, unwanted distinction: between October 1942 and August 1943 the loss rate was the highest in the Eighth Air Force (totalling more than 26,000 by the end of hostilities). Given this long and tragic history it comes as no surprise to learn that the place is haunted.

Between October 1990 and December 1991, Michael Cook was living in the RAF Officers' Mess in Keysoe Road, Thurleigh, together with one or two other Ministry of Defence civilian staff. They were working at the Defence Research Agency (formerly the Royal Aerospace Establishment) Tunnel and Airfield Sites, and with a number of RAF aircrew engaged in the research flying. The buildings dated from World War Two and were then used as administrative quarters by the 40th Combat Wing HQ and the 306th Bombardment Group of the USAAF.

On several occasions, whilst in bed, Michael heard footsteps in the corridor outside his room. When he investigated there was nobody there. He later mentioned this to other residents, to be told, 'Oh yes, didn't you know that the mess is haunted by a young American airman?'

One evening Michael was in the reading room 'When the door leading off to the bedrooms slowly opened wide (against its closed spring) and then closed again, with nobody near by, the hairs stood up on the back of my neck.'

Usually the Mess was empty and unstaffed at weekends, but some heavy snow, early in 1991, made it necessary for Michael to stay over a weekend, completely alone. Once more, he was sitting in the reading room when a light bulb fell from its fitting in the high ceiling. It dropped onto the coffee table within six inches of his right hand. 'It bounced without breaking, when I recovered from the shock, I felt a very strong sense of someone else's presence in the room.'

The Mess closed not long after Michael left and the site is now nearly derelict, apart from a memorial at the entrance to those who gave their lives fighting with the 306th. Many who flew from Thurleigh died over enemy territory, but some lost their lives much closer to home. On 23rd December 1944 two B17 (Flying Fortress) bombers collided in mid-air over Thurleigh, with no survivors. Michael has wondered many times since his encounters, whether the soul he felt and heard in the Officers' Mess was one of the airmen killed in that tragic accident, who is still trying to report back for duty.

The The well-known local dowser, Keith Paull, has been researching the history of the 306th Bombardment Group for many years. He is in contact with quite a few Americans in the States who served with the United States Air Force, including John Ziarco, the crew chief who maintained the aircraft which stood on the hardstanding which later became the car park that Keith regularly used back in the eighties, when he was working in the Electronic Training School, (part of the Royal Aircraft Establishment, based at Thurleigh). On two or three occasions, early in the morning, Keith noticed a strong smell of frying eggs and bacon when he used the car park near the school. It was very localised, being at first strong, then, on the other side of the car, as he crossed the car park, there was no trace at all. Curious, he asked his friend John Ziarco what it might mean. John had been responsible for the maintenance of 'Patches and Prayers', a B17 bomber from the 423rd Squadron of the 306th Bombardment Group. John told him that where the car park now was would have been where the ground crews had their mess tents during wartime. Once the aircraft were made ready, the ground staff retired to their tent and prepared a hearty breakfast, usually bacon and egg. The tensions suffered by the ground crews were every bit as real as those suffered by the aircrews. Would 'their' aircraft return safely? Would any of the crew be killed or injured? Would anything go wrong with the bomber? Would it be as a result of anything the ground crews themselves had overlooked? These were the kind of thoughts endured by these young American servicemen as they sat down to their morning meal. Is it possible that these stressful times somehow left an 'imprint' which manifested itself over forty years later as a ghostly aroma?

During Keith Paull's time at Thurleigh, the airfield was regularly patrolled by Ministry of Defence policemen. One of them related to Keith the story of the haunted airfield hut. On dark nights, patrols had reported seeing wartime aircrew playing cards in lighted huts, but as the policemen moved around to the door to check, the hut was once again in darkness and empty. A policeman with his dog also spotted a man on a bicycle one evening in the half-light, and then the cyclist turned off the taxiway towards a small hangar. When the policeman and his dog rounded the corner to cut off the 'intruder' there was no sign of the man, or his bicycle. It would have been impossible for any cyclist to have got to the hard-standing down the taxiway that the bicycle had taken, because a covered connecting-way between the hanger and another new building had subsequently been built across it. John Ziarco identified the cyclist as a young wartime airman who had a premonition that he would be killed. As time went by and he survived more dangerous missions, he grew more psychotic. He was still utterly convinced that his 'time was up', but successfully completed his final mission without mishap – shortly afterwards he shot himself – behind the small hanger on the hard-standing where his aircraft would have stood... These, I am sure, are just a few of the ghostly stories surrounding Thurleigh, which is currently owned by the racing-car driver, Jonathan Palmer. There is now a small museum here, open to the public, dedicated to the memory of the brave young American airmen who served with the 306th Bombardment Group. When so many young men died sudden, violent deaths, far from home, it is small wonder if the last place they knew should be revisited by their restless spirits.

Bedfordshire's most famous ghost story concerning airmen is centred on Cardington. On the fourth of October 1930, the pride of Ramsay MacDonald's Labour Government, the state-built airship R-101, slipped her moorings at 7.34 p.m., bound for Karachi – 2,390 miles away. Airships would revolutionise air travel, as unlike aeroplanes they could fly through the night, without stopping for fuel. Mooring towers had already been built at Ismailia, Egypt, at Karachi, India and at Montreal, Canada, with journey times estimated at two or three days, five or six days and three days respectively. Eventually a fleet of airships would regularly connect all corners of the Empire, including Australia.

The VIP guest list was headed by Lord Thomson of Cardington, Secretary of State for Air; the R-101 was his brainchild. His over-riding concern had been to return from a successful round-trip to India, in time to make a grand entrance at the Imperial Conference of Dominion Prime Ministers in London in October. Their support was needed for the airship program; also Thomson had his eye on the post of Viceroy of India. He insisted on the timing of the trip, despite the concerns of many of his subordinates, particularly the captain, Flight Lieutenant H.C. Irwin, who felt that the ship had not undergone sufficient trials. Sir Sefton Brancker, Director of Civil Aviation at the Air Ministry, had a blazing row with Lord Thomson just days before the fateful flight, after expressing his doubts that the R-101 was fit to go. Two Directors of Airship Development had warned Sir Sefton that the R-101 was not airworthy. Brancker was told by Thomson in no uncertain terms that the flight was going ahead, and if he was too frightened to go he should stay behind – Sir Sefton stormed out of the meeting.

It was an ominous beginning on October the fourth 1930; the ship had to drop four tons of water ballast, almost half of her total, just to gain sufficient lift to get under way. The weather deteriorated rapidly; over Hitchin the craft was pitching and rolling badly, and one aft-engine stopped in the rain and gusty winds. She had suffered problems since her maiden flight the previous year.

That same year a pilot who had died while trying to pioneer the east to west route across the Atlantic, Captain W.G.R.Hinchliffe, made contact through a séance. 'I do not want them to have the same fate that I had, as Johnston (the R-101 navigator) was a good friend of mine. I have tried to impress them myself, but it is inconceivable how dense these people are. If the flight is put off, it will be all right. I wish to goodness it were possible for you to tell Johnston in confidence, and ask him to be careful. I know what I am talking about.'

At the RAF Pageant in June 1930 the R-101 had suddenly and inexplicably dived to within 500 feet of the ground twice, from her holding altitude of 1,200 feet. Back at Cardington over sixty holes were found in her hydrogen-filled gas bags. Repairs and modifications were hurriedly carried out, including a whole new mid-section, which increased her length from 732 feet to 777 feet; the huge airship was now more than three times the size of a jumbo jet.

The Meteorological Office sent a warning on the wireless to the R-101, as about forty to fifty miles an hour winds and cloudy rain were expected over France during the passage to India. Just after nine-thirty that evening the airship left the lights of Hastings behind, for the darkness of the Channel. Henry Leech, chief engineer, while working on the faulty engine, noticed that the ship was only some 700-800 feet above the water. At ten o'clock Lieutenant Commander N.G. Atherstone, the First Officer, was forced to grab the elevator wheel, Do not let her go below 1,000 feet,' he warned the coxswain. By eleven o'clock the engine was repaired, and at 11.36 the R-101 sent a wireless message back to Croydon, to say she was crossing the French coast at Ponte de St. Quentin – the sixty mile journey from Hastings had taken them roughly two hours.

At about 2.00 a.m. the R-101 appeared over Beauvais, so low that, according to witnesses, she barely cleared the church steeple. The hull was moving broadside, not forward, and the doomed ship seemed out of control, whipped by wind and rain. The lights on the R-101 went on and off several times as she got lower and lower. Attempts were made to regain height, including dumping precious fuel, but to no avail. The airship finally hit a hummock of earth with a tremendous explosion, as the five and a half million cubic feet of explosive hydrogen in her buoyancy chambers ignited. There were two more explosions, and as the ship blazed, her mid-section collapsed, as if she had broken her back. Passengers and crew screamed for help, but forty-eight of them perished in the inferno; no officers or passengers survived, only Henry Leech and five other crew members. It was said that at the time of the crash Flight-Lieutenant Irwin's telephone extension had flashed on the Cardington switchboard, although there was no one in his office at the time.

Three days later, an Australian editor and reporter, Ian Coster, had a three o'clock appointment in South Kensington, at the National Laboratory of Psychical Research. He was there to meet the famous trance medium, Eileen Garrett, in a bid to contact Sir Arthur Conan Doyle, who had died a few months previously. The Director of the NLPR was Harry Price, a renowned psychic investigator, who was conducting the experiment, together with his secretary, Ethel Beenham. Instead of contacting the late author, the voice of Captain Irwin, killed on the R-101, came through. The following séance was to be unique in paranormal history. At tremendous speed the dead airman's voice proceeded to relate precise details of exactly what caused the R-101 to crash. Ethel Beenham's accomplished shorthand barely kept up with the rapid flow of information.

On a separate occasion Major Oliver G. Villiers, of Air Ministry Intelligence, contacted his dead boss, Sir Sefton Brancker, during a séance with Eileen Garrett. The former Director of Civil Aviation spoke to Villiers, via the medium, telling him that there had been a meeting between the senior airship officers, Lord Thomson and Sir Sefton, just before the ship left her mooring tower. The officers were so concerned about the R-101's problems that they wanted to postpone the voyage. Lord Thomson felt that they had to go, as the public expected it; tragically when Sir Sefton was asked his opinion, he felt that he shouldn't show faint-heartedness and agreed to go.

It was only later that the welter of facts received from Flight Lieutenant Irwin could be verified by Will Charlton, the Supply Officer to the R-101. There were some forty references to technical information, and twelve other reference points in total, which couldn't possibly be known by Eileen Garrett, Harry Price, Ian Coster or Ethel Beenham, at the earlier séance. Irwin mentioned 'almost scraping the roofs of Achy', which Harry Price couldn't even find on ordinary maps. It is a small village, which was shown on special large-scale flying maps. 'Strakes' were spoken about, which nobody had heard of before, but they turned out to be a naval expression (Irwin had been in the Navy), later used in connection with airships, and referred to longitudinal plates running parallel in successive strata to form the sides of the ship. The S.L.8 was talked about, but only after checking German records was it discovered that it stood for the airship designer Shutte Lanz, the eighth ship in the series. Most remarkable was a reference to the mixture of hydrogen and oil, then a secret experiment, not yet put into practical use, and known only to a handful of technicians.

Long before the ill-fated final journey of the R-101 bad omens were the talk of Cardington village. Following the disaster, many relatives of the dead airmen came forward to tell about warnings, usually in dreams, that meant many who had flown on the airship had done so with a marked reluctance. Sir Sefton Brancker, Director of Civil Aviation, probably sensed his impending doom. He had received an unequivocal warning, a week or so previously. Eileen Garrett was gripped by 'a psychic impregnation of certainty' – she was by now convinced that the R-101 would crash. She felt it was her duty to warn somebody in authority about the impending disaster, and she had her opportunity at a party, where a mutual friend introduced her to Sir Sefton. She tried her hardest to dissuade him from flying, and he also had an earlier warning than anyone else when his horoscope was cast, years earlier; it left the years following 1930 an ominous blank. Others had premonitions of a different kind. Mr. G.W. Hunt, the chief coxswain, had told his family that he feared he might never return. One engineer was so afraid to join the flight that he rode off on his motorcycle and was killed instantly when he crashed into a lorry. As Walter Radcliffe, one of the riggers, left home for the voyage, his young son suddenly cried out: 'I haven't got a Daddy!'

I will leave the final word to Captain Hinchliffe, whose predictions, from another world beyond ours, went unheeded. The day after the R-101 crash his message came via another séance, witnessed by Emilie, his widow, 'I know that death is not the end, but I hold life on earth as important to progress as life here, and wilful disregard of warnings is suicide.' Not surprisingly, to this day the R101's old hanger at Cardington is regarded by many as an unlucky spot.

Chapter 4

Evil at Clophill

No account of haunted Bedfordshire would be complete without some mention of the strange events surrounding the village of Clophill. It is a long linear village, surrounded by woodland, in Mid-Bedfordshire. The High Street runs east of the A6/Bedford to Luton road. The village has a number of attractive old houses and some mellow stone walls. At the west end of the High Street is the nineteenth century lock-up and pound.

Clophill has become notorious for two reasons; firstly, Satanic rites practised in the church of St Mary, up on the ridge above the village, and secondly a famous murder. This site gets my vote as the most evil place in the county, with good reason. A church has stood there since Norman times; the present, derelict building was abandoned in 1830, for a larger one down in the village. Up on the slope at Deadmans Hill, the old church fell into disrepair and stories of Witchcraft, Satanism and Black Magic became associated with it. Locals avoided the place after dark as the old altar stone was reputed to be used for Black Magic rites. It was said 'You could feel the presence of evil by just being there'. St. Mary's church is also known as 'Black Magic Church' by some of the locals.

In the spring of 1963 damage was done to seven altar tombs. The story received both national press and television coverage. Vandals had tried to dislodge the heavy stone slabs, but discovered that the entrances to six of the tombs were sealed up with bricks. However, they must have been fairly determined, for they eventually broke into the grave of Jenny Humberstone, who died in 1770, aged 22. She had been the wife of the othecary, her coffin was broken into and the skeleton removed. In what appeared to be some kind of bizarre Black Magic rite, the bones were found inside the church ruins, scattered around in a circle. Nearby cockerel's feathers were discovered and the skull was found impaled on an iron spike in the church porch, amidst a 'magic circle' of human bones. Strange symbols were daubed on the surrounding walls in white paint. Some represented the human eye. It was thought that Necromancy rituals had been invoked, utilizing a skeleton as a mediumistic force. Necromancy is the ancient black art of communicating with the dead.

In March 1993 the site was visited by a television crew from Associated Rediffusion, in the company of the well known writer on ghosts and witchcraft, Eric Maple, for the popular This Week programme. The unfortunate man was instructed to make his commentary from the tomb, among the bones, and a ladder was placed in the desecrated grave. Things did not go according to the programme makers' plans however, and poor Mr Maple fainted. 'All the time we were there I had the horrible feeling that something or somebody was watching over us,' he commented afterwards. 'There was an atmosphere I can only describe as absolutely evil and I never want to go there again.' The vicar removed the remains to a new location. When the vandals returned they found the skeleton gone, and proceeded to smash up what was left of the coffin, which was discovered scattered about the church and churchyard. Who knows what further evil rites had been thwarted? Eight tons of earth finally protected Jenny's last resting place, when she was reburied in her former grave by the church porch.

In 1969 a different tomb was damaged in an apparently similar attempt to remove a body. The vicar kept vigil for a couple of nights but on the third night, when the graveyard was unprotected, two more graves were vandalised. In 1975 the desecrators were back. Remains were removed from a tomb and scattered about. A visiting reporter discovered a skull on the ground by the defiled grave, together with a statuette of the Virgin Mary with the head missing.

Tony Broughall, a local ghost hunter, visited the site with his wife to take some photographs. As they were completely alone when they took the pictures, they were amazed to find that in one shot there appeared a white figure by the south window of the ruin facing down the nave. 'It was all the more puzzling because the floor of the church is at least six feet below the bottom of the window,' he commented. He felt that the figure could have been the apparition of a clergyman standing where the pulpit had once been. He later met a schoolmaster from Dunstable who had received permission from the church authorities to undertake a dig in the church interior with a small group of senior pupils. They unearthed a cross bound with reeds, where the former altar would have been. At the tower end they found more evidence of Black Magic ritual – a doll, daubed with mysterious symbols. In the nave they dug up a tomb containing two skeletons, together with a coffin nameplate with the name Sophia Mendham who died in 1893. It transpired that some of the older residents long believed that the ruins of St Mary were haunted by 'Sophie's ghost'.

There is a grim legend surrounding the graveyard of the old church at Clophill. A certain tombstone, if stared at long enough, will pronounce the date of your own death!

During the First World War two friends, who grew up together and were always close, joined The Royal Flying Corps (forerunner of the RAF). In 1914 George volunteered his services as an air mechanic, while William became a pilot. One day, returning from a mission, William's aircraft, obviously in trouble, staggered into view. George ran to help his stricken friend but William's craft hit the ground and burst into flames. George got mortally injured trying to save his pal and died soon afterwards. One of the restless spirits said to haunt the churchyard at St Mary is reputed to be George, endlessly moving from grave to grave in a vain attempt to find his missing friend's epitaph.

The second cause for notoriety in Clophill village – evil but not ghostly - was a murder which shook the British nation back in 1961, periodically appeared in the media quite regularly, and resurfaced in the national press again in 2002. This was the infamous 'A6 murder'. Briefly, the facts are these. At about 9 pm on the evening of the 22nd August of that year, a young married man, Michael Gregson, and his girlfriend, Valerie ˜torie, were in Gregson's Morris Minor in a cornfield at Dorney Reach. 'denly a man, brandishing a Smith & Wesson revolver, got into the back seat and the young couple's nightmare had begun. For over four hours they were forced to drive around the countryside at the stranger's whim, on an aimless route. Finally, around one a.m. he ordered them to stop, so he could have a sleep. Gregson drew into a lay-by on the A6 at the top of Deadmans Hill, Clophill. It turned out to be a prophetic name for poor Gregson. The abductor tied Valerie Storie's wrists together and without warning cold–bloodedly murdered Michael Gregson with a shot to the back of the head. Valerie Storie's long night of terror was far from over. The killer then raped the defenceless woman and forced her to pull her lover's body out of the car. As she lay helpless on the ground, her attacker callously fired at her several times, seriously injuring her legs. The girl managed to keep perfectly still and the murderer, believing her dead, drove off. A passing farm worker discovered her some five hours later at approximately six forty- five a.m.

The evidence surrounding the case became extremely muddled. Two suspects were interviewed by police. One was Peter Louis Alphon and the other was James Hanratty. Both men had aliases, both had criminal connections. First Alphon then Hanratty became prime suspect. People changed their evidence and Hanratty altered his alibi. Valerie Storie even changed her description of the murderer! At the identity parade Miss Storie failed to pick out Alphon. At another I.D. parade she identified Hanratty by his way of saying f for th, as in 'Be quiet I'm finking', a phrase heard on the night of the murder. Two ex-con acquaintances of Hanratty repeated alleged conversations with him which pointed to his guilt and his second alibi seemed unsound – he was charged with Gregson's murder. At the time the 21-day trial made history as it was Britain's longest. On the 17th February 1962 Hanratty was found guilty at Bedford Assizes. On the 4th April he was hanged, despite an appeal and a 90,000 signature petition.

Doubts lingered in many people's minds. There was the muddled evidence against Hanratty, some of it from known criminals. Subsequently Alphon privately confessed to crime reporters, claiming that the gun went off accidentally, killing Gregson. He promised to tell all at a press conference in 1971, when a book about the case was due to be launched but withdrew at the last moment. There were no further confessions from Alphon. This highly controversial case was not to be laid to rest until 2002, when Hanratty's guilt was unequivocally established, with the comparison of Hanratty's DNA samples and samples taken from Valerie Storie's clothing at the scene of the crime.

As for the derelict ruin of St Mary, high on 'the lonely hill' the ghost stories are not yet laid to rest. It continues to cast its spell over those of us who are fascinated by the paranormal. Allegedly, in 2001, an enterprising duo from a local newspaper decided to maintain a vigil by the church. They chatted, companionably, around the campfire that they had made, eventually succumbing to sleep. Something rudely awakened them around 8a.m., but they saw nothing and eventually made their way home. It wasn't until the photographer developed his film that the shock set in. There were six photographs, not taken by either of the two men, that showed them sleeping peacefully on the ground – surrounded by a number of other, ghostly white figures sleeping around them! This was (supposedly) one story that didn't make the newspaper and the photographer was no longer earning his living from taking pictures. True or false, fact or fiction? It neatly illustrates the point that fakes are sometimes confused with facts, and my source had been convinced by the veracity of the story. It is a great tale but, alas, it is just that, a tale, told to entertain. My golden rule when investigating supposedly true ghost stories has always been 'If it's too good to be true it isn't true'. I seemed to recall a magazine article a couple of years back with this interesting, but invented, tale in it and I think that my informant may have read the same story. There were enough clues for you weren't there? Let us take a closer look at the 'facts' – what self-respecting newspapermen would fall asleep on the case if there was the possibility of a decent story in the offing? Why would he need to develop his photographs – digital cameras have been used by the press now for many years. If such photographs existed they would have been worth a lot of money, so why were they never published, indeed why did the newspaper never use this story? Wasn't it convenient that the photographer was no longer earning his living from taking pictures, thereby making him even harder to trace? I did some checking with friends in the local media and, not surprisingly they didn't know anything about this supposedly real-life incident. Further patient and careful research revealed the truth – this Clophill incident was based on a work of fiction by Fliss Evans, a freelance writer writing under the pseudonym of 'Felix Vickery' in the now defunct Ad Hoc free weekly entertainment magazine (26th October-2nd November '02 issue). The article was entitled 'Haunted' and was a spoof story made up for Halloween.

The ruined church on Deadmans Hill continues to act as a magnet, both for people with a genuine interest in the paranormal, and for the merely curious, jokers and thrill-seekers. Although in recent times there have been no reports of further Satanic activity, the site is a popular 'attraction', particularly for the young. On Halloween groups of Bedfordshire youths regularly gather here in the hopes of some psychic occurrence, or just to try and frighten each other.

I took no chances when I visited St. Mary's for the first time on a dank night in late November 2002; I took five friends (Sue, Denny, Julie, Barry and Paul) from APIS (Anglia Paranormal Investigation Society) with me. The isolation of the place is what struck me first. In the middle of Clophill village our small convoy turned up the long, narrow, unmade road that winds its way, steadily ascending through the woods, with only our headlights to guide us, up to the lonely church on the hill. It immediately struck me what an ideal spot this had made for those practitioners of the Dark Arts, who had thoroughly desecrated the place back in the early 60s. Screened by trees, in complete darkness, isolated from any neighbours, and well off the main roads, this location was perfect for the most debased of Satanic rites. It would be a simple matter to post look-outs along the track and if necessary to turn away unwelcome visitors. When we switched off our car engines and lights the darkness was profound. We were equipped with torches and used them to illuminate the ruined building that towered over us, roofless and forbidding. Two of our number, Julie Watts and Denny Lock, were practising mediums and they worked well as a team on that night.   Denny, in particular, immediately had extremely strong feelings about the site and would not have ventured further unaccompanied; she urged us to keep close together, which we did. When we entered the shell of the church Denny was certain that Black Magic rites had been practised here. She actually felt quite nauseous; such was the aura of evil that the girls asked us to form a defensive circle, then they carried out a protective ritual to avoid 'attachment' to any of us by entities with harmful intent. Eventually we made our way outside and slowly circled the church. Barry Waddington had already told me that there was one corner of the building that felt 'different to the rest', but he would say no more. Once we had made our circuit I correctly identified it; on a cold night this particular corner was, somehow, colder than the rest, and it did have a different 'feel', difficult to articulate but nevertheless real enough. Barry told me that he has tried this 'experiment' on numerous occasions and every time the new visitor has identified the same part of the ruin that I did. Meanwhile Julie and Denny continued to receive impressions; together they pointed to a spot inside the church and both agreed that sacrifices had been made there. This had once been a Pagan site of some significance, it seems, long before the old church had been built. The most interesting encounter happened in the same area, when Julie and Denny contacted the spirit of a Victorian girl, who had been murdered and whose spirit wandered here. Later Barry confirmed this as a most convincing experience; unknown to the mediums, Barry had visited Clophill some time ago with a girlfriend who was psychically developed, and she too, had described meeting the spirit of a murdered girl. The three ladies in our group, Sue, Julie and Denny, all felt that we had 'outstayed our welcome' and were eager to leave, so we made our way back to the cars. Julie turned to face back the way we had come, as she felt the presence of a man trying to follow us who was ill-intentioned. Our group united to oppose him with a few well-chosen words; we didn't want to take any unwelcome visitors home with us.

Clophill's reputation for hauntings and strange happenings remains undimmed. In recent years visitors to the ruined church continue to make remarkably similar and consistent statements, about an 'overwhelming sense of being watched', 'as if someone was standing next to me, waiting to show me the way out', of 'something watching me leave', with feelings of 'oppression', of being 'very uncomfortable', there is 'such an atmosphere', and it is 'so forbidding'. Two ghosts that regularly feature in eyewitness accounts are a 'white woman', usually seen at the empty window, hovering about five feet off the ground, and a monk, who is seen in the churchyard, or occasionally on horseback, on the pathway leading up to the site.

I had expected that, after nearly forty years since the Satanic desecrations took place, the aura of evil that surrounds the church of St.

Mary's, like a fog that surrounds the coast, would have gradually dispersed. My nocturnal visit there leads me to believe otherwise. Thankfully, coming face-to-face with evil is a rare event in most people's lives, and at the many haunted sites I have investigated this is one of only two instances where I have genuinely felt myself to be in an evil place. The other site is also in Bedfordshire and features in a later chapter. As for Clophill's ruined church, my best advice is that it is to be avoided; I am of the opinion that it has more than likely witnessed at least one murder, the celebration of Black Mass and possibly even human sacrifice in its long history. These things can certainly leave bad psychic impressions behind. If your curiosity gets the better of you, don't go there alone, don't go at night, especially in the winter months, and don't linger there. You have been warned!

Chapter 5

The Higgins Hauntings

Haunted lavatories are more commonly reported than you might imagine, particularly in buildings which are used by the general public. This may cause you some unease when next you answer a call of nature in an unfamiliar pub or hotel, but ghosts are as likely to make their presences felt in 'the smallest room' as in any other part of a building! A haunted lavatory is just one of many strange aspects of a well-known Bedford Town landmark. I was being shown around one day, by the helpful and knowledgeable Jenny Clarke, Secretary to the Cecil Higgins Art Gallery & Museum for the past eighteen years. During my personal guided tour, we stopped in the ladies loo. It felt odd, to say the least, to be interviewing someone in such a place! Jenny herself had an eerie experience here one day, while washing her hands; suddenly the door behind her slid shut of its own accord. I tried the door myself, but it was heavy and took some force to slide shut. The ladies' toilet doubles as a lavatory for the disabled. It was a hot August day when I visited, yet it seemed to grow colder in the room as we talked, and Jenny showed me where a staircase had been bricked up and the brickwork now formed one of the toilet's walls.

Our tour began with the Gallery, which usually houses watercolours and lace. Two years ago there was an exhibit featuring several dummies in period dress. The curator was the last to leave the building and the first to return in the morning. She was amazed to find her ladder, which had been left by a screen, now blocking the fire door several feet away. Not only that but the dummies, which had been closely grouped in a display, were now all standing in a straight line along the gallery. This was a typical example of poltergeist activity.

I was deeply interested in finding out the history of the Cecil Higgins Art Gallery and there is an awful lot of history! The site of Cecil Higgins' house has a history that goes way back – to around 1100 – and Castle Close is so named after the Castle of Bedford. The Mansion itself is situated where the Castle Mound would have been. This ancient site had a violent history during the time of King Richard; England was a troubled land, King John, Richard's brother, allowed anarchy and wickedness to flourish. William de Beauchamp, traditional champion of Bedford, fell on hard times and was dispossessed of Bedford Castle. The castle was captured by Sir Fulke (or Fawkes) de Breaute, a soldier of fortune and ally of King John. Fawkes was a hated, ambitious, murderous tyrant, who led a raiding party on St. Albans, ransacked the town and attacked the Abbey, whose bailiff was brutally hacked down. So powerful had the de Breautes become, that King John's successor Henry III came to Bedford in1224 to deal with the threat. By then Sir Fulke had escaped to Wales, leaving his younger brother, William to defend the castle, it was a considerable campaign. Eventually the outer defences were captured and miners removed the foundations of the Old Tower; while giant catapults battered the walls and archers poured a lethal hail of arrows into the defenders, tunnels beneath the Old Tower were set alight and its walls collapsed. On the 14th August 1224 the garrison finally surrendered to Lord Lisle of Rougemont. William de Breaute and 73 men were arrested and hanged outside the castle walls, though three de Breaute knights pledged themselves to the Crusades and were cut down while still alive. John de Standen was ordered by the King to destroy the castle so that it could never again be a source of trouble.

But what of Cecil Higgins himself, founder of the Gallery? With Jennie Clarke's help I was able to study the life of the man and to learn about the home that he grew up in. The present Victorian Mansion in Castle Close was constructed in 1846, and was described in those days as a 'First Class' residence. It had a south-facing garden, with a strip of land that ran down to the River Great Ouse. The Mansion was built by Charles Higgins, for his son George, Charles was a distant relative of the Higginses of Turvey, and in 1823 he had become the tenant of the Swan Inn, Bedford. By 1837 he rented Castle Close from the Duke of Bedford in order to build a brewery and a house there. When he died in 1862, he was a wealthy man, owner of 800 acres of farmland; as well as being a successful Anglican businessman he was also a staunch Liberal and held the posts of both Mayor of Bedford and Chairman of the Board of Health.

George Higgins (Cecil's father) was born in Wellingborough; when he was six years old the family moved to Bedford and he was educated at Bedford Grammar School. By the age of twenty-one he was a junior partner in the brewery and took control in 1847. George continued the family traditions of Anglicism and Liberalism, becoming a J.P., a member of the Town Council, and an Alderman. A man of many parts, he was a strong churchman, and although hot tempered he was a great host; his house was always full of people with widely opposing views, he was a passionate huntsman and followed the Oakley Hunt. He married Caroline Colburn, a Dubliner, in 1844; she seems to have brought much of the fine furniture, china, clocks, books and plate which still fill the house. The couple had four children, the youngest being Cecil, the older children being George, Lawrence and Edith; in 1851 a nurse and under-nurse looked after young George and Lawrence. There were also four other servants; in 1861 a French Governess was in employment there too.

Cecil's mother died when he was eleven, and by the age of fourteen he had a step-mother – Sophia – whom he got on well with. By 1833 both Cecil's parents were dead and the house was no longer full of visitors. Lawrence, Edith and Cecil continued to live at Castle Close and to run the brewery. In 1884 Lawrence and Cecil leased the Mound, where the keep of Bedford Castle used to stand, from the Duke of Bedford, and the family built a summer-house here. Thought to be one of the finest bowling greens in England, the lawn on the Mound was used by the Higginses for croquet and tennis, while an herbaceous garden was created where the Gallery Extension now stands. Cecil was later to visit Ceylon, South Africa and Marrakesh; by 1887 he had moved to his own house at Newnham Lodge in Newnham Street, but a few years later he quit Bedford for London and bought a house near Berkeley Square in the West End.

What kind of man was Cecil Higgins? He was described as tall, imposing and autocratic but kindly, a bored yet hospitable man-about-town, who moved easily amongst the smart moneyed set. He had excellent taste and an eye for quality, and enjoyed filling his house with beautiful things. As his collection increased it became his ambition to set up a museum in his name, in his home town, which would be 'for the benefit, interest and education of the inhabitants of, and visitors to, Bedford'. Cecil was fond of riding in stylish cars (first a Mercedes and then a Rolls Royce) and he had a uniformed chauffeur; in 1935 his new car was specially adapted so that he could wear a top hat while sitting in the back seat!

Lawrence retired in 1907 and Cecil took over the management of the brewery in Bedford while continuing to live in London. The following year Cecil and Lawrence bought the lease of the castle mound and the garden round it from the Duke of Bedford. In 1909 the brothers offered it with Castle Close and its garden to the Corporation for £12,000, but the offer was refused. A year later Castle Close was sold to a Bedford doctor, who then sold it on to Bedford Corporation in 1924. In 1919 Cecil moved for the last time, to 9 Queen Anne Street in the West End. When Lawrence died in 1930 Cecil offered Bedford Corporation the use of his collection if they supplied the building and paid for staff. His capital would be left to Trustees, enabling them to continue purchasing for the collection; in this way the bulk of his fortune would go to setting up the museum. Cecil's later years continued to be active, as he had a good social life, gave dinner parties and played bridge, whilst he remained on the Bench and continued to visit Bedford for meetings. In his eighties he was still shrewd and competent, generally regarded as amazing for his years; when he was eighty-one he was visited by Queen Mary, who came to see his wonderful collection of china at his Queen Anne Street home. Cecil Charles Norman Colburne Higgins died aged eighty-five at Exmouth on the ninth of April 1941, and his ashes were scattered on the family grave in Bedford cemetery.

We continued our tour and entered the small Victorian Mansion. The Drawing Room was fully furnished and seemed tranquil and lived-in. The residents are several, but all of the spirit world. Anglia Television once carried out a vigil here in darkness and it was turning out to be a disappointing night; nothing had been seen or heard until someone remarked on this and all the lights in the watercolour gallery came on. The only way this could happen was if the mains switch, which was in a closed cupboard, had been flicked down, so with the aid of torches the cupboard was located and opened. Sure enough the switch was down, but with the return of the lights all the security guards' walkie-talkies' batteries were found to be drained! Jenny showed me a point in the drawing room where local dowser Keith Paull had located a junction of two ley lines, which are alignments and patterns of powerful, invisible earth energy said to connect various sacred sites, such as churches, temples, stone circles, megaliths, holy wells, burial sites, and other locations of spiritual or magical importance. Ghosts and other supernatural occurrences seem to be associated with ley lines; traditionally they are also paths of psychic activity, of apparitions and spirits of the dead. Could it be – I wondered – that a poltergeist had tapped into two power sources in order to move a switch behind a closed cupboard? Presumably, if ley lines provide psychic power, two of them crossing might increase that power. The stored energy, mysteriously drained from the walkie-talkie batteries might have provided a second power source.

Further mysteries were close by, at the staircase. This was not originally part of the house but a later addition. One Monday, a few years ago, while the gallery was closed to the public, Jenny was walking past the bottom of the staircase when, out of the corner of her eye, she saw a gentleman standing in a dark suit. She passed the time of day with him and continued on her way, but thought it odd that he had not acknowledged her. When she got to the staff kitchen she realised that the only two male members of staff in the building were in the kitchen – nobody was at the other end of the building! At the top of the staircase a medium sensed the presence of a 'dark gentleman' and thought that he could be George Higgins, Cecil's father. She felt that he 'toured the house and seemed to station himself at the top of the stairs'. Some nine months prior to my investigation, a young boy, on a tour of the house, got most agitated at this point; he had felt uncomfortable and was reportedly red in the face. Another young visitor reported (in September 2000) a 'tingling' sensation on touching the banister rail; I recalled that wood is alleged to be a particularly good conductor of psychic energy.

I was shown a photograph, taken in 1999, that was particularly interesting and it is reproduced in the section of photographs in this book. It shows three staff members, in period costume, at the top of the stairs. Some inexplicable lights curl through the banister rails and around the group's heads. One of the women appears in both profile and full faced! The camera and film were carefully examined and no fault could be found. A paranormal investigator (who was also a psychic medium) preceded me a few years ago, and she associated the stairs with someone who suffered with stomach pains. A current member of staff, on ascending the stairs one day, felt a male presence regarding him over the top of the banister.

There is a lovely old seat downstairs near the stairs; a medium declared that it was home to an extremely refined lady and that the seat didn't go with a middle-class home, it seems as though, as well as buying in a seat, another ghost was 'imported' too. Moving along into the Library area, the ghost of a 'stable boy', in a flat cap, (who, again didn't go with the house, according to the same medium) has been seen; he would appear to be of an earlier period and looks, in a curious way, at the books there; could he also be an 'imported ghost'? Did he come with the books? The Library may also be haunted by a man, as a previous Curator saw the reflection of a gentleman in the cabinet mirror there, but when she turned back again it was gone; a visiting printer was entering the room when he saw the figure of a man sitting at the table, but when he looked back it had gone. A member of the security staff was surprised one day in May two years ago in the hallway, when she saw a figure disappear around the corner just ahead of her, towards the stairs, only to find nobody there when she rounded the corner. The Curator added that earlier this day, in the same area, she felt the atmosphere charged and particularly cold.

Upstairs are the really 'creepy' rooms. The Guest's Bedroom has become notorious, as it is haunted by a woman, who is, according to one medium, in a frenzy of grief. Dowser Keith Paull, was 'hurled back by an invisible force' when he approached the bedroom window, and he was reported as saying 'Alright I know you are there.' Such is the atmosphere in the Guest Bedroom that many of the staff, including a former Curator, cleaners, and security men reported feeling really uncomfortable, not liking to work upstairs and becoming badly frightened when alone in this room. In 1998 a school group was visiting, and a parent who accompanied them saw a female apparition, which she later described in some detail. Her son, also in the group, saw it too, although they didn't mention it to each other at the time. After the tour she asked her son if he had seen anything in the rooms, and he gave a description which matched her own sighting – a young woman with white lace trim on her blue dress, whose face was clear, and both witnesses felt that her name was Catherine. The following year a psychic investigator confessed to feeling great unease; the dressing table mirror projected a bad feeling, so much so that she couldn't bear to look into it (Keith Paull also felt an energy force near the dressing table). Interestingly, a current member of staff who cleans the room hates this dressing table mirror. The investigator strongly felt that someone with bad eyesight had waited to die in this room. A further twist was given to the tale by an elderly lady some years ago; she was particularly interested in the Guest's Bedroom and stated, quite categorically, that a young girl had jumped to her death from a window in this room. Jenny told me that the Guest's Bedroom was due to be exorcized by a local medium.

The Nursery is another 'hot spot'. Jenny's first uncanny experience at the house occurred in this room, and she remembers it vividly. She was laying down a toy train track when she felt an almost overwhelming desire to get out. Visitors have witnessed cupboards closing by themselves, while at the beginning of last year a cuckoo clock's pendulum fell off, but instead of falling straight down or sideways, as expected, it shot outwards several feet away from the wall. An amateur dowser was frightened when she felt herself squeezed hard as she attempted to dowse the room; a young visitor, while walking through, saw the pages of the music on the piano turning over. The 1999 investigation revealed the following information about a room which is full of vibration and tapping noises. In the left-hand corner of the window sits a young woman with light-coloured hair and head bowed: the investigator got the distinct impression of being watched, pains in her stomach indicating that the occupant suffered a great deal from stomach problems. Most recently, some six months before my visit, a visitor was hit in the solar plexus by an invisible assailant, leaving her winded and shaky; a security guard, on following her into the room, felt the 'backlash' of the force pass right through her body!

The Smallest Bedroom, off the Costume Room, is one that many people find cold at all times of the year, even with the heating on (it is not open to the public). There is a tiny hatch here which leads up to the loft; in the early 1980s some workmen were carrying out electrical repairs there one day, but almost immediately they came down again. (I would not like to have been in a scramble to be first down through such a narrow aperture). White-faced, they explained that someone else would have to do the job as they had no intentions of ever setting foot up there again. In 1999 a paranormal investigator stated that this room had a 'different' feeling to the rest of the house; she actually hated it here and suffered a deep feeling of unease. This part of the house is on what was Bedford Castle Mound; there was a strong impression of a distressed boy imprisoned in a wooden tower during these times, and a visiting medium this year 'saw a man in the corner, who had been hanged here centuries ago'.

We continued on into an odd-shaped room – the Hexagonal Gallery or Militaria, which is not a particularly welcoming room, having no windows and a cold stone floor. In 1998 a visiting teacher witnessed a bowl of hot oil move across the table, teeter on the edge, then rock back and forth. A medium, unaware of the room's military connections, saw a soldier walk through the wall, she also saw a Druid, holding a large book.

My guided tour ended in the General Office where, Jenny informed me, files disappear, only to reappear at some later time neatly restored to their proper place. It was here, in 1994, that a portrait of Cecil Higgins fell off the wall despite the string being intact on the back and the nail being in place. In the adjoining Curator's Office the previous week two secretaries were startled by an electric typewriter that suddenly gained a life of its own; without benefit of human assistance it began to type away furiously by itself. That same year a pane of glass broke without any obvious reason; there was no brick or stone anywhere nearby. The glazier commented on the strange damage – it broke in the shape of a 'C'... In the Conservation Room that year a photocopier was temporarily installed, but one afternoon a member of staff was shocked to witness the lid flying up and the paper-drawers shooting out; as she headed for the lifts to go downstairs she sensed a rushing wind, yet all the windows were sealed. The 1999 investigator psychically felt the presence of a mischievous female in the Conservation Room and a man standing near the lift door. In the 1980s staff heard the old telephone system making 'dialling out' noises (which only happened when someone was making a phone call) in the area by the downstairs lift door, in the Back Hall; it was heard by several different people, always at night, always when they were alone. During this same period security staff would regularly hear footsteps – 'hurrying and scurrying' on the back stone stairs, but nothing was ever seen – the stairs/passageway were blocked off in 1974...

At last I made my way out of the Gallery, having spent the most absorbing hour with Jenny. I read in a brochure that that the 1999 'Phantom' (Society for Paranormal Investigation) had certified the Gallery a 'Grade 2 Haunted Building'. The sheer numbers and variety of phenomena in one house impressed me. Glancing at my catalogue, I discovered that even in this Entrance Hall there had been strange occurrences; only last summer, while the Gallery was closed and the shop counter unattended, a passing security guard saw the chair behind the shop swinging as if somebody had just got off it. Months later, shortly before Christmas, while two members of staff talked with the person manning the shop, they felt a cold draught float past them. The person behind the counter saw, at the same time, on the television-monitoring screen, a mist coming down to where the two stood; the rewound tape recorded no such mist.

Jenny told me that, when the Gallery was under any kind of threat of closure, manifestations increased. Happily the future of this lovely old mansion looks fairly secure and the tourists continue to trickle through steadily. Should this change I have no doubt that either Cecil or George will make their presence felt.

POSTSCRIPT: During the rewrite of this second edition I asked Jenny Clarke to keep me informed of any developments at the Gallery, and she wrote back to me in early May '04 to add a few new strange occurrences. Generally it had been quiet for a while, but then, on a weekend in February '04, there were some different odd events, beginning on Friday 13th, with an abnormally strong smell of perfume at the bottom of the stairs, which remained for about an hour, concentrated in one particular spot, and everyone who passed through smelt it.

The following Sunday, the 15th February, again in the downstairs Mansion, a member of the security staff was on duty looking towards the window of the Drawing Room when she felt a strong pull on her jacket – she spun around but there was nobody about. That same day another member of staff reported an incident from just inside the Watercolour Gallery by the Baring Room doors. As she stood there she felt a sharp tap on her legs just below the knee – she looked down but the episode could not be explained. Was the earlier happening evidence of a phantom lady, and the latter, mischievous episodes, the work of a ghostly child I wondered?

There was an even more exciting development, which I was particularly pleased to learn about, which involved someone who is not a part of the Cecil Higgins staff, someone I know well, and someone who is a most careful and accurate observer – my dowser friend Keith Paull. I will quote directly from his report, written on the 7th October 2004:

'I thought it might be a good idea if I set down some kind of report on the ghost sighting on Friday the 24th September (2004) just gone.

It happened at approximately 9.00pm just after that evening's Ghost Tour had ended and you (Jennie Clarke) were shepherding the members of the public through the Watercolour Gallery and out to the Main Entrance. June (Paull), Victoria, Geoff and myself were a little behind the main party but following on. A few paces after passing through the glass door from the main house into the Gallery we became aware of a fairly strong localised smell described variously as smoky bacon or as camp-fire wood smoke. We discussed and enjoyed the smell for a short while then Geoff dashed on through the Gallery to catch you up and maybe bring some folk back to experience the smell for yourselves.

This left June, Victoria and myself standing in a roughly equilateral triangular formation and from my position I could see the glass door into the main house over Victoria's shoulder. It was then that I saw for a brief moment the figure of a man dressed in dark clothing moving across from right to left behind the glass. For the moment I assumed, as that evening Geoff was dressed in a black shirt and trousers, that it was he, acting as 'sweeper-up' to make sure that no visitors were left behind. Hardly had that thought entered my head when I remembered that he had in fact gone on ahead of us to tell you of the smoky bacon! As soon as this realisation dawned I dashed out through the door into the hallway but, of course, there was no-one there.

So, was this another appearance by the gentleman in a black suit that has been reported several times in the area? Hopefully this report will help swell your records and at least it is firmly tied down to date and time.'

This record is an attempt to catalogue the multiple mysterious occurrences and impressions reported from over a dozen different locations, by a wide variety of witnesses, male and female, young and old; all of them have been experienced first-hand in Cecil Higgins' former home. There is increasing evidence which indicates that the unseen presences that haunt the Cecil Higgins Art Gallery remain here still. I have agreed with Jenny to carry out a full investigation, including an overnight stay at the Gallery, later this year, with some of my friends from Anglia Paranormal Investigation Society. We feel honoured to be the organisation chosen for what should develop into one of the most interesting pieces of research that we have undertaken so far, and all of us are eagerly anticipating a most exciting project.

Chapter 6

Biggleswade Bogies

One of the areas of Bedfordshire where I struggled to find any interesting ghost stories was Biggleswade. It wasn't till February 2003, a couple of months after completing the first edition of this book, that I got the break I needed, with the help of my friend Donna Regis, who was then living in the town, she put me onto the ghost of 'Aggie', who haunts the Pound Stretcher premises in the town centre. Nobody seems sure how the ghost got her nickname, but it seems to have stuck. One chilly afternoon in mid-February I sought out my Biggleswade ghost, at 115-115a High Street and interviewed Sue, who had been an assistant at Pound Stretcher for two years and had lived in Biggleswade for over twenty years. Sue told me that the present shop used to be Cavendish Furniture, and gave me a tour of this interesting old building, which she believed to be one of the oldest in the town. The cellars, used as stockrooms, have two bricked-up tunnels: one led to the church and the other to what is now the bus station – all who work here find the cellar a creepy place. Above the shop are the offices where the ghost has been known to close the sash windows after staff have opened them on hot days and she shuts the doors too, when no-one is about. 'Aggie's other favourite trick is to switch on the light in the gents' toilet which, as there were no male staff working there at the time, was not used. We looked up in the spacious attic room, empty but for some odds and ends of stock. Noises are heard from here, as of somebody moving about, but on investigation nobody is ever found. Unaccountable footsteps tread the staircase when everybody is working downstairs. I talked with another young staff member who had heard stories of a ghost-child that was supposed to haunt the premises. While working on the till one day the young assistant felt 'somebody' pass by but the far end of the shop turned out to be empty. Something strange is definitely going on at number 115. All the girls I spoke to who worked here confirmed that they often arrived first thing in the morning to discover stock littered all over the floor, and Sue has even had the experience of goods falling on her head, although they were safely stored and showed no signs of moving until she was standing below them. This was witnessed by other workers.

The general feeling was that 'Aggie' gets angry for some reason, which results in these phenomena; maybe it is her way of objecting to the alterations to this fine old house which has turned it into a bargain-hunter's delight – certainly the staff here have got more than they bargained for! I learned from Donna that shortly after my visit all hell had broken out at the shop with things scattered right and left from the shelves by an unseen presence...

I had left my job in Bedford to work in Hertfordshire when I heard from another long-term resident of Biggleswade. Roz Thompson recalled that 142 London Road, where I worked for a time, had formerly been known as 'The Old Limes'. It had been a workhouse, which later became an extension to Biggleswade Hospital, for geriatric patients. 'The Old Limes' was demolished in the early 1960s and the site became the Department of Social Services' offices at 142 London Road.

During Roz Thompson's younger days her mother, Brenda Gillard, was a nurse at 'The Old Limes', where the ghost of a woman reputedly walked the hallways. On many night shifts Brenda heard the opening and closing of doors, followed by footsteps, in areas where all the doors were locked. Invariably, when one of the elderly patients died, the noises would be heard, which led some to conclude that 'The Grim Reaper' had visited in the night and announced his (or her) presence...

There are other old legends surrounding Biggleswade, like those about Rose Lane School, some fifty or sixty years ago, when children were scared to go near the place at night for fear of encountering a famous 'headless body', said to lie in wait by the bridge, crossing it after dark required quite a lot of courage. 'Murder Bridge' on the Potton Road is another landmark declared by some older Biggleswade residents to be haunted. The story goes that a nurse was murdered many years ago and the murderer dumped her body underneath the bridge, which passes over a shallow brook.

Just to the north of Biggleswade, a short hop up the busy A1 is the site of a local mansion that gained notoriety in the last century, called Tempsford Hall. It was reported back in 1844 that Mrs. Elliott, wife of the owner, shortly after moving in, had had heard footsteps crossing the hall from the dining room to the study, followed by the sounds of bolts being opened and closed. The house owners immediately made a careful search of every room, but could find no-one. Their servant, William Johnson Butler, swore on oath about his odd experiences at Tempsford Hall (before it burned down in 1898). He heard numerous, varied and inexplicable noises; such as scratching on the panels of his bedroom walls and a loud banging that sounded like his doorpost had been hit by a mallet; also his bedroom door would fly open and then slam shut, all of which terrified the unfortunate man…

Biggleswade also boasts a haunted pub, The Golden Pheasant, which was mentioned in The Comet in October 2003. The landlords reported hearing footsteps above their heads when nobody was upstairs at the time, and the ghost, who is nicknamed 'Maurice' by the locals, has allegedly been caught on film. There may even be two ghosts, because phantom singing was heard in the bar – but it was reputedly a woman's voice!

Team APIS (Vicki O'Dell)
Some of the merry APIS crew on an investigation.
Left to right: Liz, John, Andy, Joan, Emma, Damien, Carol and Barry.

38 Mill Street, Bedford (Mark Head)
A Bedford town landmark and site of a well-documented haunting.

St Mary's Church ('Black Magic Church') (Damien O'Dell)
Clophill's notorious ruin. A brooding presence, an isolated position and still an active site for Satanism.

Woburn Abbey (Damien O'Dell)
This is actually the front of the great house, originally built in a square. The east wing, which would have been facing the camera, was demolished in 1950, thereby making the building u-shaped. Numerous unhappy spirits have been reported here.

Cecil Higgins Museum Interior (Copyright Linda Parker)
This photograph, taken in 1999, shows an unexplained light running up the haunted staircase
and around two of the ladies who were posing in Victorian costumes.

Fairfield (Damien O'Dell)
This former Victorian mental hospital, site of multiple hauntings, is a listed building. This photograph was taken during its conversion into luxury flats – hence the protective fencing. Note the broken windows caused by vandals.

Chicksands Priory (Photograph courtesy of Bedfordshire County Council)
Bedfordshire's oldest building (over 850 years of age). It has also, undoubtedly, been one of Bedfordshire's most haunted.

Waterside Mill, Barton-le-Clay (Damien O'Dell)
The Barton area is a centre for various mysterious occurrences, particularly crop circles. A tranquil setting for a tragic story.

Houghton House (Damien O'Dell)
The Countess of Pembroke's former Elizabethan home, supposedly the model for John Bunyan's 'House Beautiful'. Still in the care of ghostly guardians?

Screen 4, UGC Cinema, Aspects Leisure Complex (Mark Head)
Mark and Ryan, Anglia Paranormal Investigation Society (APIS) researchers, in row A at the haunted theatre, but which one is in seat number eight?

Cardington (Mark Head)
The hangars at Cardington, on the left is the hanger that once housed the ill-fated R-101 airship.

Priory House, Dunstable (Mark Head)
This 13th century building played an important role in both the town's history and the history of England, particularly in the annulment of Henry VIII's marriage to Catherine of Aragon. APIS' overnight investigation also revealed that it was haunted.

Flitwick Manor (Damien O'Dell)
An extremely well-documented haunted house, but a comfortable and charming place to stay nonetheless.

Chris Robinson
The 'dream detective' himself, Britain's paramount precognitive dreamer, and a native of Dunstable.

Dunstable Road and Peddars Lane, Stanbridge (Damien O'Dell)
This is the spot where Roy Fulton unwittingly gave a lift to a ghost!

White Hart Hotel, Ampthill (By kind permission of Peter Chapman and Andrew Underwood)
The haunted 'Hart' – before the disastrous fire that gutted the building in July 2001.

Wilden Manor (Damien O'Dell)
It is the track which leads up to this ancient building rather than the house itself that is subject to poltergeist activity. An elm tree, victim of Dutch elm disease, was responsible for damaging the roof rather than the lightning strike of popular myth.

Keith and June Paull (Damien O'Dell)
Keith, the dowser's dowser, at work with his pendulum in the gardens of The Five Bells at Cople.
Keith discovered the skeleton of a horse when this photo was taken. It was later discovered that
the Army used to bury dead horses here.

Pound Stretcher Shop, Biggleswade (Mark Head)
Behind the façade of a modern shop lies a building which is one of the oldest in the town, and
home to the ghost of 'Aggie'.

The Five Bells at Cople (Damien O'Dell)
Headquarters of Anglia Paranormal Investigation Society (APIS) and the retreat of some spirits not found in bottles!

Cecil Higgins Museum Exterior (Mark Head)
Bedford Town's most haunted house? Does the spirit of Cecil Higgins still watch over his former home?

BBC Radio 3 Counties (Ryan Chambers)
Mark Head, APIS member and photographer, standing outside the main entrance to the radio station's recording studios at Hastings Road, where a haunting has been reported in the reception area.

Ye Olde Plough, Bolnhurst (Mark Head)
Dating back to 1480 this ancient and atmospheric former pub plays host to the ghost of 'Cedric', believed to be a former farm worker.

Inspired! Dunstable (Damien O'Dell)
Pictured in their New Age shop Inspired! at 26 Middle Row, Dunstable, are owners Caroline and Sam. They share their shop with a number of ghosts, including that of a ghost-child called Sara.

The Bedford Arms, Leighton Buzzard (Mark Head)
Reputedly Linslade's oldest pub and formerly known as The Corbet Arms has a ghost in room 3.

Chapter 7

The Fairfield Phenomena

A short history

Fairfield Hospital changed its name in the 1960s; when it opened in 1860 it was called The Three Counties Asylum. Its progress from Victorian asylum to modern psychiatric facility illustrates the profound changes in thinking about the treatment of the mentally ill in British society. It is a sad fact of life that mental illness affects one in five of us sometime in our lifetime, so Fairfield's story is relevant to many people. The 'three counties' involved were Bedfordshire, Hertfordshire and Huntingdonshire, who responded to overcrowding in places like Bedford Asylum. Victorian thinking was paternalistic, advocating the healing power of work and moral teaching allied to the therapeutic benefits of fresh air and a rural environment. Asylums, however, lacked medical or psychiatric treatment and housed 'inmates' or 'pauper lunatics'. In these unenlightened times epilepsy was classed as a form of insanity and sufferers shared dormitories with those who had various forms of mental illness. Prior to the nineteenth century private madhouses and subscription hospitals' concern was primarily restraint, requiring the use of chains and shackles. Often this was accompanied by neglect and ill-treatment.

In 1828 the County Asylums Act meant that proper records had to be kept and regular inspections made by a Committee of Visitors. There were nine county asylums: between 1828 and 1842 another eight were erected and the next legislation passed was the 1845 Lunatics Act. The Commissioners in Lunacy were made responsible for monitoring the management of all asylums, private madhouses and hospitals for the insane; the Commissioners reduced abuses of the old system and placed the emphasis on ensuring a comfortable physical environment for patients.

In January 1856 the eminent architect, George Fowler Jones of York, was appointed to design the new asylum, to be built at Arlesey. It was of the 'corridor design' popular from the 1840s to the 1870s, with separate buildings attached to a central block by long, well-lit corridors. People did not have to go through wards or working areas when moving through the building. In March 1856 Major R. Hindley Wilkinson of Stotfold sold two hundred acres for £11,000 and the Committee of Visitors then required a further fifty acres to connect this parcel of land with the Great Northern Railway (G.N.R.), about three quarters of a mile away. In June that same year the purchase was concluded with Jonah King – fifty-seven acres for £2,892. It was to become Arlesey Tramway, used to transport coal, building materials and other essential supplies to the asylum. A further three years and nine months were to elapse before the first patients arrived from Bedford Asylum.

Builders William Webster of Boston, Lincolnshire, won the tender, at £53,626, and work started on the first of May 1857. By July it was obvious that, amazingly, not enough attention had been given to the water supplies; some 10,000 gallons a day would be needed to satisfy the demands of the new building, but initial bores were producing far less. Enter the Reverend James Clutterbuck, keen amateur geologist, who suggested a drive into the main well shaft with horizontal cuttings to provide extra well-heads. At the same time the depth of the shaft needed to be increased to seventy-three feet (twenty-two metres) – his proposals saved the day.

The Three Counties Asylum was completed in three phases, the first being the original main block of the building. The first twelve patients moved in on the 12th March, 1860. The T.C.A. now had a farm and a bailiff responsible for livestock, plus two hundred acres of land surrounding the asylum. In 1868 phase two was decided on, as overcrowding was a problem an increase of 180 patients was required with two recreation halls (the sexes were strictly segregated). Various problems delayed completion till 1871, at a cost of £22,000. Just six years later it was recognised that population increases and rising incidence of mental illness meant that capacity had to be raised to 1,000 inmates. Phase three was to provide this as well as two isolation wards (for infectious diseases) and a new chapel (St. Luke's opened in December 1879), at a cost of £72,000, which was completed in 1881. Now T.C.A. had its own farm, fire brigade, tailors, shoemakers, painters, plumbers, blacksmith, even a brewer!

Between 1910 and 1920 two new classes of patients were admitted to T.C.A., the criminally insane (in 1911) and shell-shocked soldiers (in 1917). Working conditions for staff improved with the superannuation scheme and the asylum workers' union. The First World War heralded a huge change – from passive to active treatment, largely stimulated by an attempt to treat thousands of shell-shocked soldiers.

In 1927 the T.C.A. became known as the Three Counties Hospital and 'inmates' now became 'patients'. The hospital had its own choir, theatre, band and cricket club (all were made up of staff members). During the 1920s and 30s the cricket side were recognised as a county-level team! Patients continued to work where possible – the females in the laundry or sewing-room, the males in the various available trades or on the land; while those unable to work spent long hours in the 'airing courts', large outdoor areas surrounded by high railings, a standard feature of Victorian asylums.

From 1938 to 1940 more new buildings were added to T.C.H; a female nurses' home, admission block and the hutted Emergency Hospital. During World War Two, as London hospitals were prone to bombing, the authorities looked around for rural locations where patients could be safely treated. In March 1939 the Emergency Hospital Scheme was initiated by the Ministry of Health. Parts of London's Royal Free Hospital were accommodated on site. Hundreds of T.C.H. patients were evacuated to Cambridgeshire and Northamptonshire, many from Huntingdonshire. By August 1947 those patients from Huntingdon became permanently evacuated, thus ended an association between T.C.H. and the county of Huntingdonshire which had lasted some seventy-nine years. Wartime staff shortages meant that Irish student nurses were recruited in June 1944, setting a trend that continued for some forty years.

More change was on the way with the introduction of the National Health Service in 1948, when forty-eight per cent of all hospital beds were psychiatric. The arrival of the N.H.S. swept away autonomous mental hospitals and placed them in a structure which sought to ensure uniformity of patient care, working conditions, and training for staff. The hospital farm was sold off in 1956, and by 1960 patient labour was outlawed. The 1959 Mental Health Act was a radical advance, as it confirmed a patient's right to be treated with consent where possible. The 1960s were revolutionary in terms of mental health treatment. Advances in pharmacology brought psychotropic medication to the forefront – tranquillisers, sedatives and anti-depressants. Chlorpromazine was most notable and available from 1954 – it had a powerful anti-psychotic effect. Mental hospitals, like Fairfield (as the T.C. H. became in the 60s), were now quieter places with a more settled atmosphere, and many patients were able to return home. Medication was not a cure for severe mental illness but alleviated many of the distressing symptoms and gave a greater level of control to the patients themselves. Under Enoch Powell, the Ministry of Health issued a document, A Hospital Plan for England and Wales. It laid out a policy that in future the mentally ill, mentally handicapped and the elderly would no longer occupy hospital beds for long periods. Psychiatric care was to be concentrated in the district general hospital, which would provide acute treatment in all medical specialisms. The intention was that patients would enter hospital for only a short period, for diagnosis and assessment; their ongoing care would then be a matter for community services. The writing was on the wall for Fairfield, but incredibly it was to remain open for another thirty-eight years.

There was occasional bad practice at Fairfield in the modern era; a case in point was the abuse of electro-convulsive therapy. ECT was introduced after the Second World War and used to treat depression and severe psychosis; results in some cases were spectacularly successful. A bad element among the charge nurses however, used it as a punishment on some patients, many of whom were frightened of the treatment. If patients were 'cheeky', another ploy of the few 'bad apples' among the nursing staff, was to isolate these patients and shut them in solitary confinement, in a padded cell.

With the advent of the 1980s more and more patients were transferred to care in the community, where they became known as 'clients' rather than patients. Although it has had its critics, having worked in a community support role I would have to say that in the vast majority of cases it works extremely well. Fairfield finally closed its doors in 1998. So ended the history, but what of the hauntings?

During my time in community mental health I heard stories from clients who had spent time in Fairfield, many of whom told me that it was badly haunted, and some related their experiences. The obvious answer is that my clients, having suffered with mental health problems, had delusions of ghosts, couldn't differentiate between dreams and reality, hallucinated or just imagined these things. I am rather more open-minded and believe that the truth probably lies between the two schools of thought on the matter. One says that mentally ill people are completely unreliable witnesses; the other says that if you happen to suffer with mental illness you are much more receptive to psychic phenomena. Then I started picking up stories from professional colleagues confirming that Fairfield was an extremely haunted site. One particular colleague, Jenny Butterworth, impressed me more than anyone else as an outstanding witness. She has had extensive paranormal experiences, spread over some fifteen years, while working at Fairfield. Jenny has an amazingly accurate memory for dates, times and other detail (how I envy her). She is also completely down to earth; one of the most honest and trustworthy people it has been my privilege to work alongside. Here is Jenny's story:

Jenny was a highly experienced State Enrolled Nurse who had worked at London's Maudesley hospital before joining the staff of Fairfield in 1983. She didn't have to wait long before her first ghostly encounter at Fairfield. A few months after starting, Jenny was on a day shift in the A5/A6 'sheds', in charge of two nursing assistants and some thirty geriatric patients. A 'hands on' person, she was helping her assistants, who were more used to being told what to do by their SENs than working together as a team. She moved a wheelchair from the dining area (A6), to the dormitory (A5). When she returned with a second wheelchair the first had mysteriously moved some fifteen feet away, by a partition. This happened at the start of the shift, at 7a.m. Puzzled, the nurse returned the first chair to its rightful position near the second chair. The next time she entered the room, shortly afterwards, both wheelchairs had been moved to different locations, without a sound being made.   On questioning her assistants Jenny was told that she would 'soon get used to this sort of thing happening'. Jenny was certain that her assistants weren't playing tricks and all the patients were asleep in their cots. She was, later on, to learn the truth of that statement.   Another early encounter that Jenny clearly recalled involved a phantom seen in broad daylight. In the days before Jenny could afford a car, she was walking up the long sweeping driveway to the hospital when she saw one of the domestic staff ahead of her, in a distinctive green uniform. When next she looked the figure had completely disappeared; it was when she compared notes with another staff member that Jenny realized that she had seen one of Fairfield's regular apparitions.

The Haunted Reception

Jenny's career progressed and she was soon made up to Senior SEN and transferred to Rehabilitation. One summer evening in 1993 a patient took an overdose which meant that the Senior had to take her to the Lister Hospital at Stevenage, where the patient had her stomach pumped. It was two in the morning by the time the nurse returned to Fairfield, while waiting at reception Jenny saw the window slide open, but nobody was behind the desk. Eventually the tall figure of Fred, the night receptionist, appeared and Jenny explained that the window had slid open, apparently on its own, while nobody appeared to be about. 'Didn't you see him?' Fred queried. 'See who?' Jenny replied. 'Why, the ghost of course.' Fred was perfectly serious, 'Leave your keys there and he'll put 'em away for you,' he continued earnestly. Just then the phone rang and Fred left the front desk to answer it.   Jenny did as requested and placed the keys on the counter. She looked away for a few seconds – and sure enough the keys were gone; when Fred came back from dealing with the phone call the nurse explained what had happened. Fred checked the rack, and the keys were hanging right where they should be, and Jenny felt goose bumps. Fred was known as a serious man, 'There's always someone sitting there', he told her, 'some people see him, others don't.' Jenny recalled that she had always experienced a 'cold feeling' by the reception.

The M10 Room

The M10 building was part of the Rehab Centre. One particular room, opposite the cupboards, remained empty most of the time, for good reason, although it was a bedroom. Jenny recalled a winter's night in 1990 when she was on night duty with a male colleague. All the patients were fast asleep and it was the quiet part of the shift. Suddenly there was the sound of heavy, dragging footsteps, as though someone where trying to walk while shackled, coming from the empty bedroom. The two night duty staff looked at each other; neither wanted to go out and investigate, but duty meant that they had to check. The male nurse had 'an eerie feeling' and didn't want to open the door. Reluctantly he opened the bedroom, which was, as expected, quite empty.

The footsteps were a regular occurrence, around one in the morning, on night shifts. Frankie, the ward manager, was an extremely cynical character, but he always advised against putting a patient in the haunted room, unless it was a dire emergency. Even he didn't like this particular room – nobody liked it. Sadly, the patients that did use this bedroom either relapsed into violent behaviour or, in one particular case, committed suicide.

Poltergeists in B1

Jenny's first night duty on Ward B1 turned out to be a memorable one, on a winter's night in 1996; once the patients were all settled part of the routine was to lock all windows and doors. This particular night Jenny was confronted by an anxious security guard, who told her that the fire door and the windows were open. On checking, the nurse found that the guard was correct, so she locked them all again. Later that night the security man was back – with the same story, so the door and windows were refastened for a second time; there was no rational explanation as to how the door and windows could possibly unlock on their own, and not just once but twice!

The Dark Stranger

During the fifteen years that Jenny served at Fairfield she only ever encountered two apparitions. The second encounter happened while Jenny was chatting with Christine, a nursing officer, in the grounds outside 'A' hut corridor. It was about one-thirty in the morning and as they looked over towards the hedges they saw a small, male figure in a heavy coat. They looked at each other and wondered if it was Peter, one of their colleagues, but decided it wasn't. When they turned back to have another look the figure was gone. There was nowhere for it to hide and no time for it to have moved out of sight.

A Haunted Kitchen

Mary would never forget her first night on the geriatric ward; the nurses were all gathered in the nurses' station, and in the next door hut stood the kitchen. Mary volunteered to make the hot drinks, but when she got to the kitchen she realized that she had forgotten one of the nurses' orders. It was policy to keep the kitchen door shut, so Mary called through the hatch 'Did you want tea or coffee?' to nurse B., and a voice replied 'Tea, with two sugars,' quite clearly. Mary was startled when nurse B. appeared shortly afterwards and told her that she wanted 'coffee with one sugar'. Nurse B. had heard Mary calling from next door, but had been unable to make out the words, so made her way to the kitchen, and she was certain that she hadn't passed anyone on the way. On a couple of occasions Mary was in the kitchen alone when 'someone' tried the door handle, but when Mary opened the door there was nobody there.

The Patients Who Weren't There

During our lengthy discussions two other strange occurrences were recalled by Jenny. On a night shift at ward B1, a 'patient' called out, 'Nurse, can I have a bedpan?' This was odd in itself because this particular patient was somewhat disfigured and didn't usually talk; on checking, the patient was found to be asleep. 'A3' was a mixed geriatric ward; one night while working there Jenny heard a patient call out, and this was followed by a hefty thump, as though the poor patient had fallen out of bed. The nurse rushed to investigate – only to discover an empty bed. The Nursing Sister told Jenny that this was a nightly occurrence on this ward.

Nikki's story

In 1997 Nikki was a young ambulance driver working a night shift, but on her first visit to Fairfield she had a most disturbing experience. She parked her ambulance outside Fairfield Hospital, close to the entrance for ease of access, and went to the ward for a transfer assessment of a patient, bringing with her a wheelchair. She soon realised that a stretcher was required instead, so returned to her vehicle, and was shocked to find that, instead of being close to the front doors, the ambulance was now some fifteen feet away. The car park was perfectly level, she had taken the keys with her, and the vehicle's handbrake was on, so how did a three ton ambulance move itself? She was still 'freaked' by her brush with the paranormal as she recounted this to me some five years after the incident.

Fairfield Now

I visited the huge site at Fairfield, to find that it had been divided up among several developers.   Some 650 new houses were being constructed and the main building, the 'big house', which is a Grade II listed building, was going to be totally refurbished by P. J. Livesey of Manchester. It stood there proudly, empty and forlorn, with a fence and security cameras guarding it from the vandals, who managed to do some damage before the fence was erected. Rumours continue to surround the place, but details are hard to come by, and witnesses have been reluctant to talk about their experiences.

The transformation of the 'big house' was well under way when I returned in mid-October 2004. The show apartments were most impressive, with their high-ceilinged rooms, solidly-built walls and huge windows. The Victorians, after all, had built this place to last and I don't think you would hear your neighbours with walls this thick. There were beautiful views all around, of graceful trees and manicured lawns. The kitchens and bathrooms were luxuriously appointed; furniture and fittings were of the highest quality. I was seriously tempted, even with a price tag of £250,000 for a two bed flat on a 999 year lease, but it would not suit my present needs, perhaps in years to come, who knows? I am sure that all the apartments will sell easily, to retired couples and young professional people without families, who have a taste for elegant living. However, when all the new apartments are sold, and all the new houses have been finally built, on the site that used to be Fairfield Hospital, I believe that it is most unlikely that we will have heard the last from the ghosts...

Author's note:

Many people assume, and some books erroneously refer to the 'fact' that Fairfield was, and is, in Arlesey. For the record it is actually in Stotfold.

Chapter 8

Leighton Buzzard Legends

My extensive researches have turned up little in the way of reported psychic phenomena in and around the Leighton Buzzard area of Bedfordshire. I felt, however, that it was deserving of a chapter in this second edition, as it had received no mention at all first time around. I eventually managed to find a few interesting anecdotes, after concluding that this particular corner of the county is either not very haunted or local folk are shyer than most about discussing their paranormal experiences.

One case that did attract brief notoriety happened back in 1963. Sid Mularney was a motorcycle dealer, an expert in racing motorcycles and he had his workshop in Lake Street, Leighton Buzzard. One day he decided to remove a partition to create more room. Little did Sid realise that by this innocent action he was about to stir up a virtual 'psychic storm'. Trouble began the day after the alterations were made. Upon arrival at his workshop Sid was greeted by the sight of a local racing rider's three motorcycles, their fairings smashed, lying on the floor. It seemed highly unlikely, but one bike just might have toppled over and brought the others down with it.

A few days afterwards Mr Mularney was carrying out an urgent gearbox repair and decided to stay at work until his task was finished. It was 3 am before the job was completed and he was just getting ready to go home when he felt something rush past him. He turned around to see spanners flying off their hooks on the wall, a tarpaulin, which had been covering a bike, suddenly becoming airborne and nuts, bolts and various other parts being hurled about by an invisible force. He was later quoted in the local press, 'You would have to see it to believe it. I was scared stiff. I got out of the room as fast as I could and made straight for home. My wife was asleep and I woke her up to tell her about it.'

On another occasion he discovered an enormous box of nuts and bolts, too heavy for any normal man to lift, up-ended with its contents scattered all over the floor. At other times various large items, not easily mislaid, went missing and petrol tanks were moved about. Next door was a restaurant, and the owner's small son heard loud noises coming from the empty workshop in the middle of the night, then the owner herself began to hear 'strange banging and clattering', also late at night after Mr Mularney had long since gone home.

A few weeks later the activity died down as mysteriously as it had started. It bore all the hallmarks of a classic poltergeist case, and the publicity it received stirred the memory of a lady who recalled a strange, rambling house which had stood on that site, with a large cellar that would have run under the present motorcycle workshop. At one time the site also housed a basket-making factory and local legend had it that a man hanged himself there.

The Lake Street workshop was demolished and Sid Mularney moved to new premises. The site was needed for rebuilding and road improvements, but the sudden outburst of intense poltergeist activity here remained unexplained.

Another unsolved mystery, this time more recent, from 1988, involved a photograph taken by a young man of a scenic view at Sewell Cutting; on the disused Dunstable-Leighton Buzzard railway line (it was later to become a nature reserve). When the film was developed the image of an attractive woman, wearing a 1930s style hat, framed by a railway carriage window, appeared on the print. The puzzled photographer and his family were certain that no such picture had ever been taken and the photographic laboratory was equally sure that no stray print from another film had got transferred. Did the photographer capture an image from the past, or did he manage, quite accidentally, to photograph a ghost?

Leighton Buzzard has its own haunted pub – The Bedford Arms, reputedly Linslade's oldest pub. Formerly known as The Corbet Arms (after the family who were Lords of the Manor here since the 15th century) the inn's licence dates back to the 19th century. Room three is the haunted one, and paranormal happenings have been noted as occurring between 2 and 3 am. Guests in room three have been disturbed by a lady leaning over them. She is described as; wearing a long grey dress and white pinafore, with her brown hair worn in a tight bun – many locals claim to have seen this apparition.

Paranormal researchers, investigating The Bedford Arms as recently as February 2003, turned up some interesting results. Sounds, in room three, of a woman's dress swishing along the carpet, were clearly heard when the group's video tape was replayed. In room two a woman's voice and other sounds, as of someone walking around, were also captured on tape in the empty room. In the bar area four of the party heard an unexplained loud bang from behind the bar and a cold spot was felt, although it didn't register on the temperature gauges. The entire group of eight investigators sensed that the far end of the bar was most likely to yield paranormal activity – it was later discovered that this area had, at one time, served as a temporary mortuary.

Chapter 9

Road Wraiths

I read an article in the Daily Express (Saturday October 26th 2002) about an exorcist being called in by villagers in Stoke Lacy, Herefordshire, to lay the ghost of an accident victim killed 60 years ago. It seems that twenty six crashes had occurred over an eighteen month period on the remote A465, for which the villagers blamed the ghost. Haunted highways and byways still feature quite regularly in modern ghost stories. Perhaps the most famous and also the most interesting case concerned the phantom hitchhiker of Dunstable. It was October, 1979, when a twenty-six year old carpet fitter, Roy Fulton, left a quiet Leighton Buzzard pub after an uneventful Friday night darts match. He decided to drive to another pub, The Glider, nearer his home at Dunstable. It was a dark, and foggy patches hung over the flat, open countryside. He was driving back along the Dunstable Road, coming through Stanbridge, on an unlit stretch, where Peddars Lane becomes Station Road. He saw a youth thumbing a lift, and so pulled up past him and waited for him to catch up. Roy opened the door and said, 'I'm going into Dunstable, where are you for?' The young man merely pointed towards Totternoe, a village past Stanbridge, but said nothing as he got in and sat beside the driver, so off they drove. The Minivan passed through Stanbridge and carried on down the Totternoe road at a steady 45 mph with its silent passenger. Some five minutes into the journey the road ahead was now lit by streetlamps, so Fulton slowed down and took a pack of cigarettes from his pocket. 'Cigarette?' he asked, flipping open the packet and holding it out to his companion. There was no reply, so the driver shook the packet and turned to his passenger...but there was no-one there. Roy braked hard and looked in the back of the Minivan as well – it was empty, and the road behind was dark and traffic-free. It occurred to him that nobody could have jumped from a vehicle that had been travelling at over 40mph, besides the door light would have come on if the door had been opened. It was firmly closed. His hand edged over to the passenger seat...it was still warm.

Confronted with a completely inexplicable encounter, the young carpet fitter naturally felt chilled and frightened, so he drove quickly to The Glider, where he had a large Scotch and told the startled landlord, 'I've just seen a ghost.' Later on he reported the incident to a Police Inspector Rowland, at Dunstable Police Station. This eerie tale eventually made the pages of The Sunday Express, but Roy Fuller was disappointed when nobody came forward to confirm that they too, had encountered the ghostly hitchhiker. Despite diligent searching of newspaper files from previous years, a local reporter could find no accident in the same area to explain any link with Fulton's haunting. The case has attracted considerable interest, and now features regularly in many books and magazines about the paranormal.

During the nineteen-seventies there was a rash of 'phantom hitchhiker' stories in circulation in newspapers all over the world. It became a sort of urban myth; my theory for this phenomenon is that the TV and movie actor Telly Savalas may have inadvertently begun the trend. I remember that he recalled his own encounter during an interesting interview with Michael Parkinson. The American actor, well-known for his long-running role as TV cop Kojak, was driving home from a friend's house in a remote part of Long Island, New York. It was 3am in the morning and his fuel tank was empty, but as luck would have it the lights of an all-night café led him to refuge. He went in, ordered a coffee, and asked for directions to the nearest petrol station. He was advised to take the path through the woods at the back of the café and walk until he reached a motorway.

Savalas told Parkinson, 'I was just about to set out when I heard someone ask in a high pitched voice if I wanted a lift. I turned and saw a guy in a black Cadillac. I thanked him, climbed into the passenger seat and we drove to the freeway. To my embarrassment, I had no wallet –

it must have fallen out of my pocket. But the man loaned me a dollar. I insisted I must pay him back and got him to write his name and address on a scrap of paper. His name was Harry Agannis.'

The next day Telly Savalas looked up his Good Samaritan in the phone book, and a woman answered the call. Yes, Harry Agannis was her husband, but no, it wasn't possible to speak with him, for he had been dead for three years. The actor's first reaction was shock, there must have been some mistake he felt, but he couldn't put the incident out of his mind. Eventually, he visited the woman, taking with him the piece of paper on which the stranger had written his name and address. Savalas said, 'When I showed her the paper she was obviously deeply affected and told me that without doubt it was her husband's handwriting. I described the clothes the man had worn. She said those were the same clothes Harry Agannis had been buried in.' The actor and the widow sat and looked at each other for a long time, hardly daring to admit to themselves the implications of what had happened. Was Savalas helped by the ghost of Harry Agannis? He said, 'That was a case Kojak could not solve. I doubt if I'll ever be able to explain it.'

Like many other psychical researchers I have always been extremely sceptical about 'phantom hitchhiker' stories, which all seem to contain basic similarities. I believe, however, that Roy Fuller's story is an important exception. A few years after the story appeared, an independent investigation was carried out by a psychic researcher called Michael Goss, who was most thorough. He interviewed Roy Fuller, who reiterated the events of October 12th 1979, recalling the details most vividly. His hitchhiker had been very young, about twenty, extremely pale-faced and drawn, and his face was unusually long. His hair was dark and curly; he was wearing a white shirt with an old-fashioned collar and dark trousers. A check with the local constabulary confirmed that Fuller's original statement to the police exactly matched the information he had later given to the psychic researcher. Goss also interviewed the publican at The Glider who served the badly-shaken driver with a whisky. Goss was initially sceptical, thinking that Fuller invented the tale for the sake of notoriety, but after his investigations he came to the conclusion that Roy Fuller had most definitely been telling the truth.

Pavenham has two tales of haunted roads: the first concerns a lady cyclist, on her way from Pavenham to Stevington. She was picking up speed on a downhill stretch, when she was confronted by a large white dog; such was her speed that she was totally unable to avoid the animal, then, to her relief and surprise, she found herself actually cycling through the dog!

The second Pavenham tale also happened on a hill – Carlton Hill. Another lady cyclist, this time from Harrold, dismounted to walk up the hill, and on reaching the top she saw two horses galloping at top speed towards her. She pulled her bicycle on to the grass verge out of their way. The horses were travelling so fast that the lady feared a terrible accident would occur, so she turned around and pedalled downhill, to be able to offer her assistance. She fully expected to find a crash at the cross-roads at the bottom, but there was no trace whatsoever of the horses, and the puzzled lady returned safely to Harrold. Shortly after her sighting she met another woman whose father, a Headmaster at the Reformatory, had an identical encounter – some thirty years previously!

In more recent times motorists have braked sharply on Carlton Hill when faced with a galloping phantom horse, complete with rider. The ancient sunken road which runs over the fields from Chellington towards Pavenham is also said to be haunted by the same 'galloping ghost'.

In the spring of 2002 reports were still coming in about a haunted stretch of road on the A5 opposite the Wagon and Horses pub in Dunstable, where an apparition, thought by passing motorists to be a hitch-hiker, has been seen several times. In the early 1900s a girl called May Steeples was strangled here and it may be her spirit that walks the highway.

A phantom Morris saloon was reported by Mr Stanley Prescott of Dunstable one quiet Sunday morning in spring 1961; he was out with his wife driving on the Dagnall to Edlesborough road, towards the Travellers Rest crossroads. 'I saw a black Morris saloon approaching me from the other direction, it came straight at me, and I knew if I did not take avoiding action I would be killed. My car went through a hedge into a field, and then my wife asked me how the accident happened. There was no other vehicle about, my wife had not seen the Morris, but I knew I did, it was uncanny, the whole thing was remarkable, and it frightened me.' Fearing that he may have suffered some kind of hallucination, Mr Prescott saw his doctor, who could find nothing wrong with him. There was an interesting development to the story four years later; a fatal crash took place on the same stretch of road, once again on a Sunday morning. This time a car swerved across the road, into the path of an oncoming coach. A coroner's court was unable to establish a reason for the car's sudden swerve, we shall never know if the car driver tried to avoid 'an old Morris saloon' ... only to run headlong into a coach. Mr Prescott read about the accident. He always remained adamant that the Morris had appeared real to him. His comment was 'I definitely saw a car that day. I can still remember the incident vividly. It was an old-type Morris saloon. I have often thought about it since then, and reading of the unexplained accident started me thinking again.'

An intriguing encounter was related to me by Michelle Hounsell, at her Biggleswade home. It happened back in 1997, and although it was five years ago, it was so vivid that the young mother recalled the details clearly. It was a rainy autumn evening and Michelle was being driven by her friend Andy. They passed the RSPB bird sanctuary on the A1042. This stretch of road is one I know well; it is particularly dark at night, being heavily wooded, and Andy's car had its high beams on. It was about 7p.m and the couple were on their way to Sandy Upper School. They were driving downhill, into a left-hand bend, when suddenly a figure crossed the road from the left, giving Andy no time to react. Michelle recalled being suddenly confronted, 'with a headless figure, all in white, a skinny woman with a big chest, she was in modern dress, wearing trousers, at her feet trotted a small dog, Jack Russell-sized.' The car went straight through the apparition. A shocked Michelle turned around in her seat but there was no trace of the headless dog walker – the figure had appeared so solid. It sent a cold shiver down Michelle's spine and gave Andy quite a jolt.

Back in 1958 a cricket team from Kenwood Manufacturing Company Limited were returning to Surrey, after playing at Milton Bryan, near Woburn. Their mini-bus driver swerved to overtake a car and crashed at a spot near the Packhorse Inn on the A5, near Dunstable. Two of the team were killed – Sidney Moulder and Jerry Rycham, both from Woking, and three others were badly injured. Near here, right on the borders of Hertfordshire and Bedfordshire, a taxi driver had the living daylights frightened out of him one evening in 1973. He was on his way to collect a fare when a tall man in white clothes stepped right in front of his cab. The driver said, 'I braked but was going too fast and went straight through him. I pulled up and spent some time trying to find the man but there was no-one in sight.' There have been many other reports from motorists on the road from Markyate to Dunstable, who have seen the tall figure of a cricketer at the side of the road and assumed that he had just left the Packhorse Inn. Does the ghost of one of these tragic accident victims still return to the site of his demise?

Barry Waddington, who lives in Wootton, had a strange encounter late one night in June 2002, as he drove along Fields Road, off the A421, on the way to Bedford. He came around a right-hand bend when a figure appeared without warning, forcing Barry to brake hard. As he came to a halt there was another squeal of tyres from behind him, on the same bend, and next thing he knew his car was hit in the back. Such was the impact that the following Fiesta van's airbag deployed with a loud bang. The female pedestrian, that Barry had managed to avoid hitting, asked him if he was all right, to which he replied that he was unhurt. He was surprised to notice that the woman was all in black, even her jeans were black, not the best clothing to wear when walking on an unlit country road late at night! His attention, however, was diverted by the need to check on the driver of the Pat's Pizzas van that had just crashed into him. He ran back to discover that the man was, although shaken, not injured. They both looked up to find that the woman in black had completely disappeared. It was a straight bit of road, with nowhere to hide, and it would have been obvious if she had tried to head off into the dark fields, besides there would have been no point. Barry is convinced that there just was not enough time for the woman to have got out of sight of both drivers. The other man was as bewildered as Barry, as he too had seen the woman one minute and the next minute she was gone. Barry's wife, Sam, later discussed the incident with a friend at work, who said that she had heard of similar sightings of the mysterious 'woman in black' on this particular stretch of road. My further investigations revealed that almost on a monthly basis a car ends up in a ditch on this bend. Does a 'road wraith' haunt Fields Road?

Bedfordshire certainly has plenty of haunted highways, and I am certain that there are many more such stories out there – for instance a ghostly pony and trap is one apparition that has been mentioned to me. (I have used the term 'wraiths' loosely, as an alternative to ghosts, the strict meaning of the word 'wraith' is the ghost of a person on the verge of death.)

Chapter 10

Mansion and Manor Manifestations
Goldington Hall, Flitwick Manor and Woburn Abbey

Goldington Hall

Some haunted houses have extensive recorded history, and it usually turns out to be well worth studying. If, like me, the research side of paranormal investigations holds you in rapt concentration, the Beds and Luton Archive and Records Office at Bedford Town Hall is the place to be. There is a wealth of material to be discovered about a place as old as Goldington Hall. Originally it is thought to date from Jacobean times as a stone, rather than brick construction. The oldest remaining parts, however, date from Cromwell's era. These are; some lead rainwater pipes bearing the date 1650, a few pieces of old glass that have been reworked into a design in a small upstairs window lighting the staircase on the first floor, and a patch of stone found on the north wall.

There were two tenants, Thomas Threpell and Robert Latton, whose dates of occupancy remain unknown. The first known occupier was Nicholas Luke, recorded in the Hearth Tax returns of 1671 as having no less than ten fireplaces on which he had to pay a total tax of £1 per year. The first recorded owner was Sir Thomas Allein, who sold the Hall – property and lands of Goldington as well as Ravensden, Renhold and Bedford for £4,107.0.0d to the trustees of John Davies of London in 1680. After that date nothing is known about who owned the Hall, or who lived there, until 1843, when the property was owned by Robert Faulkner and occupied by Lord Francis Russell, a relative of the Duke of Bedford.

In July 1844, William Kenworthy Browne moved in to Goldington Hall, with his new bride, Elizabeth (nee Elliott). Elizabeth's father was High Sheriff of Bedfordshire and lived nearby at Goldington Bury. William's father, Joseph, was a wealthy wine merchant and Alderman who had been Mayor of Bedford in 1842. William was active as a Magistrate and rose to the rank of Captain in the Bedford Militia. He is chiefly of interest, however, for his lifelong friendship with Edward Fitzgerald; they both studied at Trinity College, Cambridge, together with William Makepeace Thackeray. Edward Fitzgerald was a regular visitor to Goldington Hall between 1834 and 1859. He found the house cold and uncomfortable but was delighted to find that it contained furniture that once belonged to Dr Samuel Johnson. Johnson's desk was in Fitzgerald's bedroom and the poet liked to sit in the library at Johnson's bookcase.

Records show how much Fitzgerald enjoyed visiting his friends at Bedford. He mentioned them in his letters, people like Harry Boulton, gentleman farmer of Putnoe and Robert Elliott, William Browne's father-in-law. He also frequently mentioned the village of Goldington as it then was. It must indeed, have been a tranquil place; cows grazed and bridges crossed merry brooks. In modern times the brooks have been filled in, the bridges removed, the old road across the green diverted and trees planted. Goldington lost its identity as a village and was swallowed up by expanding Bedford Town. Fitzgerald was a gifted letter writer and much of his epistolary works remain to illustrate his great wit and sympathy. He was a close friend of Thackeray, author of Vanity Fair, who was another distinguished visitor to the Hall. Some of the friends Thackeray mixed with here were drawn upon to create characters for his next novel, The History of Pendennis.

After a disastrous marriage in 1856, which lasted less than a year, Fitzgerald spent more and more time seeking solace with his good friend Browne and burying himself in his most famous work – a free poetic translation of quatrains from the anonymous pamphlet – The Rubaiyat of Omar Khayam, one of the best loved works in the English language. By the end of 1857 he had sent off his completed translation of the Rubaiyat to a publisher but it wasn't actually in print until February 1859. Sadly, that same year, in January, his good friend William Kenworthy

Browne was crushed when his horse reared on him on the way back from hunting. He died from his injuries two months later.

Goldington Hall passed into the ownership of the Polhill family (probably through marriage) of Howbury Hall Renhold, who sold it in 1874 to William Marsh Harvey, barrister and historian. By now the Hall was practically a ruin. Harvey became well-known as the author of The Willey Hundreds, an extensive Bedfordshire record that took from 1866 to 1872 to complete. The new owner also set about a programme of major reconstruction and repairs at Goldington Hall. This lasted for about three years and converted the house more or less to its present form. In 1884 his mother died and new bay windows were built in her memory; when Harvey died in 1918, the property passed to his two sisters. Both were staunch Conservatives, one of them being Dame Warden of the Bedfordshire Habitation of the Primrose League. They lived here until 1924.

In 1928 the house was let to another celebrity, Mr David Robinson, founder of Robinson Rentals (later to become Granada TV Rentals), and benefactor whose philanthropy was responsible for the creation of Bedford's first indoor swimming pool. The millionaire racehorse-owner stayed in residence until 1940 (when he moved to Newmarket). Later directories list several names as tenants; they were presumably occupiers of flats (Goldington Hall had been converted some time after, when it came into the possession of Bedford Corporation). Bass Charrington purchased Goldington Hall in 1972. Goldington Hall was to become just a memory: the brewery paid £12,250, changed the property's name to The Falstaff, and converted it into a seventy-seater restaurant with four bars. The Restaurant was on the first floor with an adjoining Restaurant Bar; downstairs was the Henry IV Bar, the Bardolph Bar and the public bar – the Pistol Bar. The old front door was replaced (it was supposed to have come from Newnham Priory); traces could still be found in the cellars of underground passages which were supposed to link the Hall with the nearby church and Newnham Priory.

I began my investigations into Goldington Hall in autumn 2002, spurred on by inaccuracies in some of the local newspapers. Several publications had printed articles about 'the ghost of a mad gardener, who had attempted to kill a former owner of the Hall, and was seen staggering around the grounds with a pitchfork through his heart.' This particular legend properly belongs to The Grange, long since gone, a property which was on the other side of the road.

I called in, on October 10th 2002 at what was Goldington Hall, or as it is now known, The Lincoln Arms. It lies hidden by trees, at the corner of Church Lane overlooking Goldington Green on the north side. I was greeted by Dan Allnott, assistant manager, who gave me a brief tour of the place and a rundown on the modern history. A lot of the Hall's original land has been sold off, so that now there are just a few acres left out of an original garden of ten or twelve acres. During the Second World War the Hall was used by the intelligence services; members of the Special Operations Executive and code breakers from Bletchley Park reputedly met here for briefings. In the past thirty years the pub has had about ten different owners, the current owners being Greene King. As he lived in the building, Dan was able to confirm the rumours I had heard about its being haunted. There are two ghosts; one is a lady, an older person in Victorian dress, who haunts the window seat on the landing. I wondered if she could be Mrs Harvey; the lady whose son, William Marsh Harvey, had commemorated her after her death by the addition of new bay windows to the house back in the 1880s. She sometimes appears in the evening and there is a cold area felt around the window seat although there is no draught. The other ghost is heard but not seen; she is a little girl and 'little girl' footsteps were frequently heard (about every three days) on the upper landing, usually at around one in the morning. Dan and his friends had often heard the child walking on the floorboards and had run up only to find nothing there. Psychics who have visited the place claim that the two ghosts are happy spirits who love the old house. 'Harvey', a Golden Retriever that belonged to the landlady, often refused to go upstairs, once every few weeks he ran around in a panicky state. A previous manager had a 'fearless' Great Dane, but this dog never went in the Pistol Bar area. There are four tunnels that Dan believed were linked to the church; dead bodies of the family or their servants could be moved directly from the Hall to the church without outsiders seeing them. Three of those tunnels run from the Pistol Bar area. I found the old building quite charming, a sort of poor man's mansion, not on the same scale as many listed buildings and not as grand or ornate, but attractive nonetheless. The ceilings looked original and fairly basic compared with some of the intricate designs in grand old houses. One room's ceiling, however, was copied from those at Trinity College Cambridge, where Mr Harvey was educated. The many mirrors, an essential feature in any pub, lend it a slightly spooky air; there is plenty of solid oak panelling. Dan reckoned that the old servants' quarters up in the attics were quite creepy. I am glad that this particular house was preserved but sadly its far grander neighbours like the Queen Anne-styled Goldington Bury, and The Grange, were not so fortunate.

Flitwick Manor

Fleotwic was the Saxon name for present-day Flitwick, it meant 'dwelling on the river'; the village had originally been a Roman settlement and it continued to be centred around the high ground where the Manor now stands. Alwin was the last Saxon Thane and Lord of the Manor in the eleventh century; he was displaced by William Lovet, a Norman, in 1086. As well as the Manor the Lovet family owned, and farmed, about 500 acres, much of it woodland, with a water mill. William Lovet soon replaced the wooden Saxon church with a stone Norman one. The ownership of Flitwick passed on, first to the Saundreville family and then the Refuse family, who became Lords of the Manor in 1210: David Refuse became Sir David de Fletwycks and his descendants lived in peace at their Manor for 150 years. The family crest was a simple pair of red leopards on a black shield, quite distinctive for jousting! In 1361 the Refuse family sold their estate and there were a number of titled, City of London and Royal absentee landlords; they included the Earl of Kent, the Duke of Albemarle and King Charles I. By 1632 the Crown was short of money and sold the 'messuage' to Sir Edward Blofield for £90. Through marriage the Manor passed to the Rhodes family, who remained Lords for almost 100 years.

In 1735 Anne Fisher inherited the house on a marriage settlement with George Hesse and upon his death in 1783 married George Brooks, whose family were the Squires until 1932. Under the Brooks' management, and the Enclosure Act of 1807 the estate reached full maturity, with its long drive, beautiful water gardens, avenues of fine trees and landscaped gardens. Although there are modern housing developments all around, some extent of the former lands are evident from the Mount opposite the Manor. Here monks from the monastery, which was built close by the great house, were buried, and the dew pond they knew still remains. Today's Manor dates from the late seventeenth century, though most of the front was added in the eighteenth century and there have been further additions in recent years. Some of the lime trees bordering the drive are reputed to be 250 years old; the drive was the original road which ran between the Church and the Manor House from Flitwick to Priestley and beyond. The Manor lands contain a ha-ha and a Gothic folly.

Robert Lyall, cousin to Catherine Brooks, inherited the estate on her death in 1932. Gradually the estate has been sold off though an interest in Flitwick has been retained by the Lyalls, who lived here until the late 1950s and the house changed hands for money only three times in several hundred years. Local street names commemorate the Brooks/Lyall families – Lyall Close, Brooks Road and Catherine Road. Soon after the Second World War the Lyall family sold the Manor and after passing through several hands it was converted into a hotel in the 1960s, and in 1998 it became part of Menzies Hotels Plc.

I was introduced to Diane Rance, who has been restaurant manager at Flitwick Manor for about six years, on the 10th December 2002. We sat in the comfortable lounge, which was looking at its festive best with a roaring log fire blazing in the hearth; there was a large Christmas tree and other tasteful decorations. The antique furniture and family portraits helped to create the impression that one was a guest at a country house. On this chilly winter's day we discussed the hotel's haunted history over a pot of steaming tea. Diane had worked at the Manor for a total of fifteen years; she told me about the hotel's first owners who were staying here one Christmas some sixteen years ago. Late one afternoon the ghost of an old lady walked through the morning room as the family sat there; she appeared a little hazy but solid enough and was a traditional ghostly white. Diane has never seen the ghost herself, although one day she was startled in the kitchen when a clock fell off the wall and saucepan lids flew out of the cupboards; there was quite a racket made as the stainless steel lids hit the tiled floor. Several staff members have reported waking up in the night feeling a heavy weight on their bed and a cool hand on their forehead. The guests, too, have encountered the phantom old lady; the apparition mainly confines its activities to the upstairs bedrooms where it sits on the bed and cries. Descriptions are of a white-haired old lady who wears lace on her head; sometimes a scent of rose perfume has accompanied her manifestations.

So who is the ghost? Opinions vary; as she is benevolent and often puts her hand on people's foreheads she may have been a nurse, a maid or a housekeeper, whilst others have suggested it is Mary Brooks. What is known is that Mary and her husband John Thomas, High Sheriff of Bedfordshire, were devastated by the death of their daughter Mary Ann at the tender age of just twenty-six, in September 1848. There is also a story that a former housekeeper was fired after being accused of trying to poison an old man in the house but, after her death, returned to his room. The hauntings intensified when a small room was discovered in the roof; such rooms would normally have been allocated to servants, so the housekeeper theory is the one that is most commonly accepted.

The ghost of Flitwick Manor is most likely to resume haunting if any alterations are carried out to the building. Such was the case back in 1994/1995, when builders had to carry out essential repairs; the Manor was leaning forward and the first floor brickwork required re-modelling, some 95 pins being driven through the walls to hold the building together. During renovations the builders were working on the roof when they discovered a small hidden attic room, which was empty – the housekeeper's room? Sonia Banks, the general manager then, gave an interview to author Betty Puttick for her book, Ghosts of Bedfordshire. She told the writer that the spectre seemed to favour one particular bedroom and Sonia knew when it was about, as the cushions on the chair showed an impression as though someone had been sitting there. Sonia had plenty of other paranormal experiences, 'I have heard her walk across the floor above and slam a door when I knew I was alone in the hotel.' She continued, 'One night I was standing in the hall talking to some visitors when the lights went out, and everyone cried "Oh!" I said it's just the ghost and switched them on again, but off they went again, and I kept switching them on, and they kept going off, until I said, "Will you please stop it?" The visitors wondered who I was talking to, and I explained that I was talking to the ghost.' Sometimes locked doors were mysteriously unlocked although no one had touched them. The chef, Duncan Poyser, also told Betty Puttick that he had experienced the ghost; once on an overnight stay, he had felt a heavy weight on his legs and he couldn't move or turn over in bed. At a wedding reception, one of the musicians, a keyboard player, spotted the ghost going out of the French windows, his description exactly matching those of other, previous witnesses. Lydia Dawson, the duty manageress, was another person who had a nocturnal encounter with 'the old lady'. While staying overnight at the hotel Lydia woke up to see someone standing at the end of her bed; a little old lady with grey hair wearing a long Victorian dress and a small white cap. She appeared agitated and looked as if she was crying. Lydia thought she was trying to say something to her but there was no sound. The young girl leapt out of bed and ran out on to the landing; when she returned she discovered that the light had been switched on – by the ghost!

Staff at the time held the theory that the ghost was angry about the disruptions at the Manor. The bedroom that it favoured was temporarily roofless and was being used to store things from other rooms. A guest complained about the ghost sitting on his bed in the middle of the night and then the following night Lydia had her experience. As well as featuring in the local press the story appeared in a catering magazine, where it was picked up by researchers from the television programme, Strange but True and in April 1995 London Weekend Television filmed a re-enactment of the hotel guest's encounter.

Diane Rance told me that things had quietened down again in recent years, and that is just how the ghost likes things – quiet. Diane recalled how some eight or nine years ago a band had been playing loudly, as bands do, and when they packed up they asked Diane about the mysterious old lady who had been watching them. It seems that she definitely dislikes too much noise; on another occasion there was a loud disco in full swing when suddenly the electricity went off and stayed off for a good ten minutes! Lighting goes off and on, a favourite trick of the entity and things mysteriously go missing too. Rumour has it that an exorcism was carried out at Flitwick Manor, probably in the late nineteenth or early twentieth century but if so it doesn't seem to have been particularly successful at banishing one particularly restless revenant.....

Woburn Abbey

Woburn Abbey's story begins in 1145 when Cistercian monks from Fountains Abbey in Yorkshire, under Hugh de Bolebec, founded an abbey at Woburn. Woburn however, has a history that is recorded as far back as 969, when it was a Saxon hamlet, the name deriving from the Saxon word 'wo' meaning crooked and 'burn', a small stream. Around the Abbey the hamlet soon became a village and the village became a town, with a market three days a week. In 1538 on the Dissolution of the Monasteries, the Abbott of Woburn, Robert Hobbes, was executed for treason after speaking out against Henry VIII's annulment and remarriage to Anne Boleyn. Legend has it that the Abbott and two of his clerics were executed and left hanging from the branch of an oak tree near the south front of the house; it is claimed that the ground below the tree remains bare of grass or flowers to the present day. In 1538 the Abbey was dissolved and in 1547 the King died. That same year, under the terms of Henry VIII's will, Sir John, Baron Russell of Chenies (who acted as the Monarch's executor) was given title to several properties including Woburn, following many years loyal service to His Majesty. The Russell family originated from Dorset and it was said of Sir John Russell that he was always on the winning side, as he rose steadily through diplomatic, military and royal household duties. He had been a gentleman usher to Henry VIII, later Comptroller of the Household and then Lord Privy Seal. In1550, during Edward VI's reign, Russell gained the Earldom of Bedford. He is said to have far preferred his manor at Chenies in Hertfordshire to his Bedfordshire property; hardly surprising as Woburn Abbey was a run-down, badly neglected old monastery, just one of his staggering array of former church lands which included Tavistock Abbey in Devon and Covent Garden and Long Acre, both in London. It was Francis Russell, fourth Earl of Bedford, who built the mansion of Woburn in the seventeenth century; and turned it into a family home, when he moved out of London with his eight children to escape the plague. He also continued the Fen drainage scheme begun by his father, known as the Bedford Level. Lord William Russell was beheaded in 1683 for complicity in the Rye House Plot (the attempted assassination of Charles II and the Duke of York). During the reign of William and Mary, in 1694, the Russell family were back in favour and they were given the title of Duke. The present structure of the mansion was begun in 1744, when it was modernised by the fourth Duke of Bedford, again called John Russell, who filled the house with fine furniture, porcelain and handsome objets d'art; he commissioned many artworks, including portraits, and a number of paintings were executed for him by Canaletto. The fourth Duke was responsible for creating the park and surrounding the estate with vast quantities of trees.

There was to be much sadness in future generations of Russells, the ninth Duke committed suicide at his London home. The eleventh Duke Herbrand (who died in 1940) was married to Mary, who was an interesting and adventurous personality known as 'The Flying Duchess'. As I sat in the reception area awaiting an interview with Lavinia Wellicome, Curator of Woburn, Herbrand's portrait stared down at me; he was dressed in scarlet military uniform, and he had a long moustache; the eleventh Duke definitely had an aloof air about him. In March 1937 his wife the Duchess Mary Russell, made her last flight; her Gypsy Moth crashed while she was on a visit to the Fens; some wreckage from her aircraft was washed ashore six days later near Great Yarmouth. Hastings, the twelfth Duke of Bedford, continued the growing family tradition – of wealthy and powerful men who were yet lonely and unhappy. After many rows with his only child, notably about the latter's pacifism, father and son did not speak for twenty years; in 1953 Hastings was found shot dead on his estate in Devon. Things had been difficult for his estranged son John Robert, Marquis of Tavistock, who had lived in a boarding house in Bloomsbury until he was invalided out of the Coldstream Guards in 1940. He was a house agent, then a journalist, and later a farmer in South Africa, when, in 1953 at the age of thirty-six, he became the thirteenth Duke of Bedford. His inheritance came with strings attached; there were death duties in excess of four million pounds to be paid and the estate was burning up £150,000 a year in maintenance costs. To his horror he returned from South Africa to discover, as he said, The Abbey was in utter chaos, half of it gone and 'a great gaping crater remaining'. His father had found dry rot in part of the building – his solution was drastic, he had demolished the east wing, the riding school, a real tennis court and a museum, during 1949-1950.

The mansion's rooms were in a poor decorative state and some were crammed from floor to ceiling with furniture, pictures and china, all thrown in anyhow. The thirteenth Duke also had a radical solution to the estate's problems: he decided to open the doors of Woburn to the public, and the sooner the better. He set himself a target of just six months to become one of the first Stately Homes open to the public, and working night and day alongside his staff he had it ready. Woburn Abbey has continued to draw vast quantities of visitors, many of whom return again and again to enjoy the multitude of attractions, assuring the future of this palatial Stately Home. There is, apart from the house, a successful Wildlife Safari Park and a popular Antiques Centre. I can also personally vouch for the magnificent Sculpture Gallery, a memorable venue for private functions. Certainly the treasures of Woburn are acknowledged as one of the finest private collections in England.

Woburn also has a well-deserved reputation as one of the most famous haunted Stately Homes in Great Britain; with well over four hundred years of history, and many tragic deaths, it is home to a variety of ghosts around the mansion and in the grounds. The thirteenth Duke and Duchess were made aware of them shortly after they moved in, nearly fifty years ago. The Duchess has been quoted as saying, 'Ghosts are quite frequent and most friendly at Woburn. When I first came I did not believe in them. Now I have to believe in ghosts.' She has witnessed some strange things over the years, like the ghost that is fixated with doors; it opens doors, then 'an icy draught is felt all over your face', whilst another ghost 'gives the impression of touching your face with a wet hand in the middle of the night.' (Could this be the ghost of a drowned servant I wondered - as we shall see later a black servant of the seventh Duke died in most tragic circumstances)? One night the Duchess was walking down a corridor with her whippet Melchior, when five doors opened in front of them, causing the dog much distress. The Bedfords once owned seven dogs; when the animals sensed a paranormal presence they would cower, with their tails between their legs, and howl in the middle of the night. One of the oddest encounters the Duchess had was with thirty men and women from 'a visiting ghost club'; they were discovered chatting with the Duke in one of the bathrooms, the group pronouncing that this one room alone 'has nine ghosts.'

A particularly chilling tale concerns one of the seventh Duke's servants: burglars broke in and nearly strangled him to death, then locked him in a cupboard in the Masquerade Room. Once they had finished searching for loot they threw the servant out of the window and drowned him in one of the lakes on the estate. This black servant is alleged to haunt the mansion and may be responsible for the ghostly 'wet hands', mentioned earlier, or perhaps the opening and closing of doors... This manifestation took place soon after the fourteenth Duke took up residence: in the television room doors would open with nobody there, and then came a pause, long enough for someone to walk the length of the room, then the doors at the opposite end would open. The Duke found it distracting and draughty, but not frightening. The locks were changed, but locking the doors made no difference, so eventually the area was reconstructed and turned into a passage. Some of the guest bedrooms were similarly affected; guests discovered the communicating doors between their bedrooms and dressing rooms might open as many as five times in a night.

Monks, not unexpectedly, feature in several hauntings at Woburn, while excavation work was being carried out both a cleaner and a workman experienced a 'cold chill' before seeing a monk walk past them. During a function in the Sculpture Room in March 1971 guests saw a figure in a brown habit by the entrance pillars, which quickly disappeared through a doorway, and women have complained of being groped by invisible hands here. The Crypt, used to display porcelain, silver and gold, is the scene of monk-like apparitions, reported by workmen, visitors and staff. Could the ghost of the hanged Abbott Robert Hobbes be responsible?

In their private quarters both the thirteenth Duke and Duchess noticed a restless, uncomfortable atmosphere that they found difficult to describe. A well-known clairvoyant, Tom Corbett, once investigated the house and he found the top floor rooms in particular, 'Soaked in unhappiness over many lifetimes, leaving an overpowering atmosphere of misery, strong enough to affect people.' There is a 'malaise' felt about the Wood Library and in the Duke's office; he also felt the presence sometimes of his late grandmother 'The Flying Duchess', in the isolated little summer house on the west side of the park. It is not open to the public and it is where Mary Russell liked to be alone, to write her diary and to watch the birds.

Later ghost sightings have been focussed on the Antiques Centre where genuine 18th century shop facades have been re-erected to form imaginative shop fronts for over sixty dealers. Sometimes lights have been found on after they have been turned off and a tall man in Victorian dress has been seen to wander through the market wearing a top hat; the shade of a previous Duke of Bedford perhaps? The fourteenth Duke of Bedford was Henry Russell, who became Duke of Bedford when his father John died on the 25th October 2002, while living in the U.S.A. The fourteenth Duke, who suffered a stroke in 1987, was a rather kindly, shy man who, having lived in the Abbey for many years as Marquis, then moved with his wife Henrietta to the Dower House in Woburn village; he had three sons – Andrew, Robin and James Russell. Andrew had the title of Marquis of Tavistock, and on the death of his father Henry in 2003 Andrew became the fifteenth Duke of Bedford. Andrew currently lives at Woburn Abbey – it remains to be seen if he is as interested in the ghosts as his father was.

The Georgian village of Woburn, too, is haunted; over the years it has reverted from a town to a village. Once it boasted a population of 2,100 (in the 1851 census) with twenty-seven inns; it was an important staging post on a countrywide network that included London, Cambridge, Oxford, Nottingham and Leeds. During the 20th century the population fell to 700, though currently it is around 1,000. The Georgian style of Woburn is due to the fire of 1724 which destroyed most of the houses, necessitating a complete rebuilding programme.   It featured in the Ampthill News, in January 1974, when staff at the Bedford Arms Hotel (since renamed) were interviewed about its ghosts. The 68-year old night porter, Reg Williams, described one of the phantoms, 'He's dressed in a smock, has an old hat on and is smoking a clay pipe, about two feet long. There's a dog like a greyhound by his side. The first time I saw him was while the hotel was closed. Work was going on at the time and part of it was the opening up of the old fireplace, which had been bricked up and plastered over, I suppose about a hundred years ago.' The ghostly man and dog sit in the fireplace and have been seen on several occasions, but their identities remain unknown and they are not scary. Mr. Williams continued, 'You get a feeling that he's friendly and that you'd like to speak to him if you had the chance.' The 'white lady', however is a ghost of a different ilk, 'This was a horrible experience, it was like a load of steam coming down, I saw it take shape, and it was like a woman.' Mr. Williams had read about the burning of Woburn in the 1700s by the Duke of Bedford, who had become obsessed with an idea that Woburn was getting too big and becoming a town; (I wonder if the relentless expansion of neighbouring Milton Keynes, which threatens to engulf such places as Aspley Guise and Woburn, gives the present fifteenth Duke sleepless nights). The Bedford Arms was one of the buildings badly damaged in the fire, and a number of people are believed to have perished in the building. Certainly the 'white lady' has terrified those hotel staff unfortunate enough to see her in the reception area. Some time previously Mr. J. Graham, a hall porter at the Bedford Arms, abruptly left one night never to return and in his haste he left behind a jacket containing the key to his lodgings!

When I visited Woburn Abbey to research my story I had a most interesting meeting with Miss Lavinia Wellicome, Woburn's Curator. I could not have wished for a better guide. Lavinia had worked at the mansion for thirty years and was a charming, sensible lady who was clearly devoted to the Russell family. She made a point of the fact that the family have been close and affectionate during all her time at the Abbey, not always the case in previous generations. Lavinia joined as a volunteer in 1973 and in 1978 took up the post of Curator. Given her businesslike approach, I expected to encounter a sceptic who would relay old, well-documented stories, but instead I found a pragmatist who had first-hand experiences of her own to recount! Our meeting was held in the Crypt, an atmospheric office filled with books, files and paperwork, a cosy and warm place to work, strangely enough! Just outside in the passageway, secure in their glass cases, were a magnificent collection of fine porcelain, including Sevres, Meissen and English pieces. It was in the Crypt office during the 1980s, while working late one night that Lavinia felt 'not alone' and clearly heard the sound of heavy vellum pages in a big book slowly being turned. She thinks it was probably Mr. Pickering, who was the librarian here at the turn of the century. Also during the 1980s, a cleaning lady reported seeing a man with a plumed hat and short cape who looked like a Cavalier. In the 1960s the cook, who coincidentally was called Mrs.Cook, saw the swing door from the butler's pantry to the kitchen, swing open, but nobody came through. Ernest the butler then saw a figure in white passing through into the passageway. During this same period, when one of the buildings was being converted to a laundry, a workman fled the site saying, 'I'm not stopping here!' after seeing a figure in 'a long grey tunic' pass straight through the wall. It is thought that he saw the same phantom monk that has been reported on several occasions in the past.

Just the previous year – one day in June at around midday with the sun streaming through the shutters, Lavinia was in the Wood Library; she was concentrating on her work, sorting out various valuables, including some gold snuff boxes, when she got a strong scent of roses. She looked up and caught a glimpse of a lady wearing a deep navy blue dress of the early Victorian period, a second later the figure was gone. Lavinia talked with Derek Stanford, whose father was valet to the eleventh Duke and Derek later became a room warden, he too had smelt the rose perfume on occasion. It transpired that Georgina, who married the sixth Duke in 1803, had a favourite perfume – it was 'esprit de rose'...

Nowadays the estate in Devon has been sold, although the Russell family still have a farmhouse in the county that they use for holidays. The Covent Garden and Long Acre properties have also been sold off, though the Bloomsbury property has been retained. It seems that the Dukes of Bedford will continue to flourish in the twenty-first century and to preserve the wonderful treasures of Woburn Abbey. There are some people who believe that former church properties, which were confiscated and then redistributed by Henry VIII, hold nothing but bad luck for their later owners. Woburn Abbey is one site that seems to have more than its share of ghosts, though by no means do they all seem unhappy!

Chapter 11

Pioneer Ghost Hunter
(Bill Turner 1833 – 1929)

It was Saturday, the 21st of September 2002 and I was on a ghost hunt at Belgrave Hall; a mansion in Leicestershire made famous by programmes like Living Television's Most Haunted, where we had the best equipment a twenty-first century psychic investigator could wish for. We had motion detectors, electronic thermometers, video cameras and tape recorders inside the house. Outside, I was able to scan the gardens with night vision binoculars and was particularly impressed with these Russian-made instruments. Even the darkest corners of the extensive grounds were clearly visible as green glowing shadows. It was a successful night although I was disappointed to be the only one of our four person team not to see the ghost! It was intriguing to learn after returning from Leicester, about Bedfordshire's first serious paranormal detective – Bill Turner, an unknown, unsung hero of the supernatural. People always said there was something 'other-worldly' about Bill, who was born in 1833; he was a jobbing gardener who lived near Oakley. A simple man, Bill had a rare talent for horticulture, possessing not only green fingers but green thumbs too! He was also well known for his extensive knowledge of weather-lore and was recognised as a talented fortune-teller at village fetes. The other thing that made him stand out from the crowd was his passion for ghosts; during his long and active life Bill Turner investigated all the 'classic' Bedfordshire hauntings, such as Woburn Abbey's 'Flying Duchess' and the ghost of Willington Manor. When he died in 1929 at the grand old age of 96 his ancient deed box was discovered. Inside, along with his military medals from the Crimean War, was a leather-bound notebook. Its pages were numbered from 1 to 108, and filled with his investigative notes, hand-written with his steel-nibbed pen.

It all began when Bill's grandfather was landlord of The Old George at Silsoe, which was reckoned to be haunted. Bill never saw anything but others reported seeing a 'lovely grey lady', the ghost reputed to be Lady Elizabeth Grey. She was a member of the aristocracy – the de Grey family, whose 'seat' had been Wrest House, Silsoe, for almost 700 years. The twelfth Earl - Henry Grey, who had been Queen Anne's Chamberlain, laid out Wrest Park, after visiting Versailles. A descendant, Earl de Grey, rebuilt the magnificent house and Silsoe Church in the eighteen-hundreds. Unfortunately the beautiful, headstrong Lady Elizabeth fell in love with a man far below her station, a coachman she met at The Old George. The lovers decided that their only option was to elope, and the Lady of the Manor hid out at the inn for a fortnight while her furious father scoured the countryside. Eventually he tracked them down and the lovers got word that he was on his way. Desperate to remain together they raced off in the coach, but the young man lost control at a bend in the road and the speeding coach toppled into a lake. Elizabeth, trapped inside, was drowned, but her ghost continued to haunt The Old George, and was reportedly seen by a terrified workman, as late as1960. The previous year the landlady had advertised for 'a layer' to exorcize her troublesome ghost; it had been responsible for all kinds of annoying tricks including door-slamming in the middle of the night. As with most haunting, the passage of time brought a cessation of incidents. Many phantoms simply fade away after about a hundred or so years, and so it would seem to be with Lady Elizabeth, who has been quiet of late. Ten years ago she was occasionally 'sensed' about the inn and residents would sometimes hear footsteps.

Wrest House was not the only haunted house in the village; close to The Old George public house lived a Mrs. Hallam, and her home was to be the scene of Bill Turner's first 'sighting'. He was engaged to work on Mrs. Hallam's garden and was lodging with her. One night, after a hard day's gardening, he heard a faint tapping on a neighbouring bedroom door and then there was silence, followed by a creaking on the stairs. This happened about six times then, as Bill's hand was resting on the counterpane, he suddenly felt something brush it. He leapt out of bed and immediately lit the lamp as floorboards creaked outside his bedroom door. Wondering what horror might appear he was taken aback to be suddenly confronted with a small, fair-haired child – a girl, wearing a pinafore dress and no shoes or stockings. The distressed child ran past him only to disappear by the bed. Though he searched the room thoroughly he could find no sign of his little visitor. The next morning another lodger asked about the child. 'Who was the little girl I saw on the landing last night? She tapped on my door, and when I answered it, she just faded away into the shadows of the staircase.' Mrs Hallam, the landlady, chuckled, saying, 'Oh that'll be Sarah. About thirty years ago she lived in this house with her father and stepmother; the little girl had been looked after by a nurse to whom she was devoted, but the nurse was later dismissed and the little girl pined and died. The man who used to own this house told me that the girl quite often appeared in the old days looking for her nurse, she's quite a family friend now.'

Around 1870 Bill found himself working at a friend's large house in Ampthill, just off the main street. The place was up for offer at a low rental; it had the mouldy odour of a house that has stood empty for some time and it had a big garden, a real challenge for any gardener as it was full of weeds and couch grass. Bill stayed over for a couple of weekends to get the grounds in some sort of order. His friend told him that soon after moving in he had been in bed with his wife when the bedroom door had opened wide. The couple sensed 'something' enter the room; there then followed sounds as of someone staggering, then a gasping whisper, 'Pull it out, please pull it out.' Then there were more gasping sounds, moans of pain and finally a scream of terror. This occurred two or three times before Bill finally heard it for himself: he went to the owner, Mrs Letitia Bland, who would make no comment; eventually, using all his charm and powers of persuasion he learned the dreadful history of the house. Twenty years previously a Mrs Anne Rumbelow, a widow, and her fifteen-year-old daughter Agnes, had lived there. One night they heard someone in the garden. Agnes ran downstairs to make sure that all the doors were locked but minutes later she staggered back into her mother's room, screaming, 'Pull it out!' Mrs Rumbelow was horrified to see a sickle blade embedded in her daughter's back. Agnes died the next day, and her murderer was never found. 'Ever since', said Mrs Bland,' the house had the reputation of being haunted.'

It was on a hot summer's day that Bill Turner had one of his strangest brushes with the supernatural. It was in an ordinary parish church near Milton Ernest, where Bill entered to get out of the oppressive heat and to see the beautiful stained-glass windows. He was alone in one of the back pews when he heard the church doors open and close. Then there were footsteps walking down the aisle. Strangely though, nobody appeared. He was just about to leave when he turned around for one last look at the windows over the altar. In the front pew sat a man in old-fashioned clothes and a silk top hat. As he made his way through the churchyard, Bill spotted the sexton and enquired of him about the man in the top hat. The sexton told him that when he was a boy there had been a vicar in the parish who lost his wife suddenly. On each anniversary of her death he appeared in ghost form. 'And that,' said the sexton, 'would be to-day!'

Bill Turner's successor is probably the most famous of all modern day ghost hunters – Peter Underwood, President of The Ghost Club and author of over 40 books on the paranormal. I have a vivid recollection of seeing him at The Knights Templar School in Baldock some years ago. The school hall was darkened, the only light source being a small bright spotlight trained on the speaker. Mr Underwood kept the audience spellbound for a couple of hours with his extraordinary experiences. The strangest thing about the evening was that all the electrical equipment set up for Peter's talk failed simultaneously and much scurrying about was necessary to get the show back on the road...

This distinguished gentleman took part in his first psychic investigation in Bedfordshire. As a member of The Society for Psychical Research he was invited, along with two professional mediums, a local councillor and several other interested parties, to a number of séances at Woodfield. This was a house in Aspley Guise's Weathercock Lane. I am always on my guard when there is, or could be, a money motive linked to a so-called haunting. This is a typical story of that kind; the tale is that Woodfield stands on the site of a much older building, which was linked to a grisly local legend. Some two hundred years earlier this house was home to a young girl and her father. The young lady had a secret lover who visited when the father was away. One night however, the father unexpectedly returned and the lovers hurriedly concealed themselves in a large cupboard in the pantry. Unfortunately for them the father had already seen them through the window. He is then supposed to have locked the unfortunate pair in their hiding place by placing heavy furniture against the door and they were left there to die. None other than Dick Turpin is alleged to have happened by, as always seeking refuge from the law. He found the bodies and decided to blackmail the murdering father. From then on the house would be a place of refuge for Turpin whenever he was in the area, and the lovers' bodies were hidden under the cellar floor. Their ghosts were said to visit Woodfield regularly and horses' hooves were heard galloping down the hill. There were also reports about Turpin entering the grounds on horseback, passing through a thick hedge where there was once an entrance. In 1948 Mr Blaney Key tried to get his rates reduced because his home, Woodfield, was haunted. Mr Key alleged that the value of his house had depreciated due to its ghostly reputation. Luton Area Assessment Committee arranged an investigation. Enter Mr Underwood and company, but several weeks and a number of séances later the investigations were deemed to be inconclusive. Mr Key's case came up at the Shire Hall Bedford, after the Chairman and Counsel discussed the matter privately, and the appeal was withdrawn. The case made the headlines again a year later, with a fresh appeal to get the rates assessment reduced on Woodfield, as the house was supposedly still haunted by a ghostly horse and rider as well as a phantom lady. It all descended into farce when a witness was called who had been evacuated to the house some years earlier, during the war. She claimed that she had seen the arms of the murdered girl reaching out to her through the wall above her bed one night. When asked what she had eaten that evening her answer, 'Cheese sandwiches' reduced the court to helpless laughter and the appeal was rejected. The Chairman of the Bedfordshire Quarter Sessions Appeals Committee summed up the case as 'devoid of merit and without point or substance'. The whole story seems quite bizarre to me. I can understand somebody trying to get their rates reduced – hands up those who think they get value for money from their council tax! But did Mr Blarney, sorry, Blaney Key seriously think he would get away with it? As for the legends...what cupboard door is fitted so as to be airtight? It would have taken many days for the couple to die of thirst and starvation as they certainly wouldn't be asphyxiated. Surely they would have called out for help? Even given that the father was as hard-hearted as to ignore their pleas for days on end, would no-one else have heard them? Did they have nobody working for them and no neighbours, not even any visitors? What of Dick Turpin the ubiquitous highwayman, conveniently turning up? We are to believe that Turpin, after breaking into a house and finding two dead bodies, stayed around, woke up the master and agreed the cunning blackmail scheme with him. Extremely unlikely – so what we are left with, if we strip away the more exotic elements of the story? A possible haunting of Woodcock Lane by an unknown horseman, but then that isn't romantic is it?

I hope to live as long, as happy, as productive and eventful a life as that of both Bill Turner, pioneer ghost hunter, and Peter Underwood, king of ghost hunters in modern times. I am a traditionalist and enjoy following in the footsteps of such men.

Chapter 12

Restless Revenants of Bedford Town

The Spooky Cinema...

A modern cinema complex is not the sort of venue normally associated with ghosts. I was, therefore, intrigued as I made my way along the A421 from Kempston, one Saturday in September 2002, for my meeting with Simon Peters at his home in Wootton. Simon had e-mailed me the previous Thursday with a brief note about his experiences at the Aspects Leisure Centre in Newnham Avenue, Bedford. He had responded to an article about my investigations in the local Times & Citizen.

When I was comfortably seated Simon began his story. It all started for him back in 1991 when he was working as a 'screen host'. One night he was locking up (a procedure usually involving two staff) with Stuart Church, the assistant manager. In those days it was extremely dark in the cinema after normal hours, not as well lit as now and the pair only had their torches to guide them in screen four. Simon remembered this particular night vividly. Rocky IV was the feature and it was a wintry December evening around midnight. The two cinema employees made their way forward towards the screen, Simon was walking down the far left hand aisle and Stuart was on the right. They were intent on putting chains around the doors to secure them. The pair were about half way down when they clearly saw the left hand exit door open on its own – it was alarmed and the catches were on but the bar pushed down as though by some invisible hand and the door swung open so violently it hit the wall with a bang. Simon and Stuart 'stood there frozen in our tracks'. Simon is a young guy over six feet tall, well built and not easily frightened, but this night both men dropped their torches and ran out of the back of the room. This was to be just one of many occasions when people were spooked in screen four.

On another occasion a lady was terrified to clearly see in the mirror in the ladies toilets, the figure of a man walk past her dressed in monks' clothing. It was to prove an embarrassment for the cinema. The police were called and the woman barricaded herself in the toilet. Though the police could find no explanation, the frightened cinema-goer told them that she had seen a man's feet under the cubicle door and was afraid to come out. The ladies' toilet is alongside screen four. Other incidents occurred from time to time in the loo. Another time a little girl was heard screaming in there and she complained to her mother about 'the man in the mirror'.

Some five years later the cinema was under the ownership of MGM (originally it was part of the Cannon group). During Mr. Wright's management things in screen four were getting much worse. Complaints from the cinema's patrons were becoming a regular occurrence. Some people heard scratching noises above their heads while others felt phantom hands on their legs. Seat number eight in 'A' row was always down, yet it should have automatically flipped up when unoccupied. If staff moved it into the upright position it would be down again when they returned. One memorable evening when screen four was full, Mr. Wright himself was standing at the back of the haunted theatre when he distinctly heard chains being jingled...

Maurice, a security guard, was patrolling the complex one night, as he walked past the Apex he saw a figure in the foyer. The bewildered guard couldn't believe his eyes when he witnessed the 'person' continue straight through the glass doors and out into the car park. He later described it as a 'hooded, cloaked figure'.

There was one regular customer who visited screen four on average three or four times a week. She was an older lady who remembered the days when Newnham Swimming Pool had been on this site. She had worked there and remembered that the lockers had been haunted. They opened on their own and then slammed shut, always at the same time of night.

Whatever the weather, screen four was always colder than the other screens. Staff used to joke that if anyone felt a bit overcome with the heat in the summer they should be moved there. Temperatures remained a constant 76 degrees Fahrenheit in the other five theatres at the Apex, but in screen four the temperature never rose above 35 - 40 degrees Fahrenheit. The other thing that staff noted was that the psychic phenomena were seasonal. They happened mostly from December up till about April time. Occasionally in summer, when it was hot outside, a presence was felt. The strange occurrences were also fairly consistent in terms of time. Things usually, but not always, kicked off around 11.30 p.m. and carried on till about 7.30 a.m., when the cleaners arrived. The cleaners made sure that screen four was done first! They still didn't escape the attentions of the spook. Is it a coincidence that four is considered the most unlucky of numbers by the Chinese? It is held in superstitious dread and is to be avoided at all costs...

The cinema was empty one night and some of the staff attempted to contact the spirit with the aid of a makeshift ouija board, but the glass they were using flew off the table and smashed to pieces against a wall, frightening everyone. They abandoned any further attempts to hold a séance.

Simon left Aspects some fourteen months before our interview and UGC took over ownership of the cinema, but the phenomena continued to plague staff and customers alike. On a bright sunny day in early October 2002 I was able to follow up on Simon's intriguing revelations. I arranged to meet Joe Fatibene, one of the staff who was still currently working at the theatre, and he was able to corroborate Simon's story. The owners of the cinema have never liked talk about the ghost and past managers have forbidden their staff to spread word of the haunting, but in recent times there seems to be a more relaxed atmosphere about it all. Some workers however, have felt a bit foolish about reporting their experiences. In common with many others who have worked at the cinema, Joe Fatibene has always felt a presence in screen four. He knows the ghost as 'Raymond'. Many years ago an usherette who worked with Joe confirmed that her nephew had hanged himself from a tree in the grounds of the old priory, where screen four now stands. He was only twenty-one and his name was Raymond.

During his time at the cinema Joe has witnessed many strange things. One night a kettle switched itself back on and boiled again, despite having been switched off and unplugged! Cleaners have regularly been hit by apples and sweets while working in screen four. One unfortunate cleaner spent an hour and a half one night cleaning the place on his own. When he returned rubbish had mysteriously reappeared from nowhere. Another time, after the theatre had been thoroughly cleaned, an apple appeared on the floor in the middle of the aisle, in perfect condition but for a single bite taken out of it. It seems as if the ghost of screen four has a sense of humour! At other times cleaners, like Elaine, have been hit by flying Coke cans and apples in an otherwise empty room. Once an electric scrubbing machine suddenly switched itself off, then back on again while the floors were being cleaned. Joe Fatibene even talks with 'Raymond' every morning. He feels that in this way the ghost won't harm him. I accompanied Joe to screen four, which is a small theatre. I was immediately struck by the chilliness in the air. It was cold enough to raise goose bumps on my arms. This was at ten in the morning. Outside it was a warm 17 degrees centigrade, according to my car's temperature gauge. We then stepped into screen one, the largest theatre, which if anything I would have expected to be colder. It was appreciably warmer.

An ancient wall runs around the back of the Aspects Leisure Complex. My researches revealed that when the cinema was built, SDC who carried out the excavation work, found quite a number of bodies and it would appear that the site was originally a burial ground. In the seventeenth century (around 1625) a fire destroyed what had once been Newnham Priory.

I certainly would not be put off watching a movie in screen four at the Aspects Leisure Complex, as the ghostly experiences seem to be confined mostly to members of the staff. I am, however, convinced that this is one unlikely, but genuinely haunted, site in Bedfordshire.

Some Other Restless Revenants...

As we can see from earlier chapters, many of the haunted sites in Bedfordshire are in the County Town of Bedford. With a haunted cinema, museum and art gallery, plus several haunted pubs, the place is teeming with restless spirits, both old and new. There are quite a few more though, as we shall learn...

I personally investigated one such site with great interest, after discovering that the ghost had been seen by a colleague of mine. Jadran is a big Bosnian guy, well over six feet tall and with the frame of a night club bouncer. It would be fair to say that he isn't afraid of anything – or he wasn't until he slept at Sisters House. Situated on Saint Peter's Street, right next door to Saint Luke's Church, Sisters House is a listed building. It is rumoured that a secret passageway once linked the church to the house, which is a former nunnery that now belongs to the Pilgrims Housing Association. It is a large building on three floors, divided into eight flats that provide supported accommodation living for clients with mental health problems. Some years ago it was a rehabilitation unit for people with chronic mental health problems and staff would occasionally stay overnight while the unit was being set up. It was on such an evening back in 1997 that Jadran had his experience. He was suddenly awakened by a weight at the bottom of his bed and he realised that 'somebody' was sitting on his legs. He jumped up and switched on the light, just in time to see the figure of a nun slowly drifting silently down the room to disappear through the wall! It was the first and last psychic experience that he ever had; he flatly refused ever to spend another night on duty in that room in flat six. I spoke with other staff members at Sisters House. One admitted that things often go missing and she attributed it to the ghost and would mentally say, 'Oh stop being silly and give it back,' and usually the items would rapidly reappear. I looked around myself one warm July day, and found the place to be musty and a touch chilly, despite the pleasant weather outside. I could certainly imagine it to be a haunted place.

I have already mentioned Keith Paull, my dowser friend. Together with his wife, June, he had an eerie encounter in the middle of Bedford just two years ago. The couple were driving from Saint Peter's Street; they were waiting at a red light, opposite the Alphonse Sandwich Bar. As they pulled away from the lights they were startled to see, a few yards ahead, crossing from the pavement on their left, a man in World War One army uniform. Quite regardless of the heavy traffic he continued to look straight ahead and somehow managed to pass between the cars. He was hatless, with short hair and he crossed the road before just fading into thin air.

In 1975 number 7 Battison Street gained a brief notoriety in the local press.The Bedfordshire Journal reported that Mary Dorney had unwelcome, unearthly visitors. Having lived at number seven for less than a year she was witness to several sightings of a ghostly couple. The man favoured plus fours, while the woman wore a floral nightdress. The pair enjoyed a stroll at night in a back room of the house. During the daytime they seem to have been responsible for turning off the television when no-one was in the room. Neighbours and friends commented on the cold, clammy atmosphere in the haunted area... Mrs Dorney eventually moved to another address.

Mill Street is undoubtedly one of the most infamously haunted streets in Bedford Town. Number thirty-eight served, until 1964, as the offices of the Clerk to the Justices before he and his staff removed to St. Paul's Square. Built in 1760 the house is an ancient monument, and over the years there has been much well-documented testimony to its haunting. Mr. T. B. Porter (now deceased) of West End, Pavenham was quoted in the Bedfordshire Times (14th January 1966). He was a former town clerk and local solicitor who had an office in Harpur Street. Mr. Porter was brought up at thirty-eight Mill Street and wrote about it in a letter to the Bedford Record. 'The sound of unexplained footsteps at thirty-eight Mill Street is no unusual event. My father, a man of strong will, refused to have the matter mentioned, saying that the subject was quite unfit for discussion by educated people. As the orders approximated closely to those of the Medes and Persians, this was quite sufficient. In later years my mother did occasionally mention the matter. As small boys we used to come in to dinner at one o'clock, passing the kitchen door and going upstairs to tidy up. Mother then served the dinner and took it into the dining-room; on one occasion we seemed to be a long time and she called us to come down but got no answer. Then we actually came in.

"Have you been in before?" she asked; we had not and wondered why she had asked. She went back into the kitchen and told the old servant. "You're not going up alone," said the servant, snatching up a knife," I'm coming too." 'They searched the house from top to bottom, but found nobody. Both my mother and the servant were convinced that they had heard us come in, and could not understand it. They carried out my father's orders and told us nothing about it at the time.'

'Another experience my mother told me about happened one winter's night. She was sitting in the dining-room when she heard two heavy knocks on an old oak back door which had a heavy iron knocker. Screwing up her courage she opened it – but no one was there. She called but got no answer. When my father came back at nine o'clock (he had been at the Town and Country Club to play billiards), she told him what had happened. He said it could be simply explained by the fact that the big gates facing Mill Street could not have been bolted. Upon investigation however, he found that they had been bolted, so he then said that there must be somebody in the garden – this was about a quarter of an acre and well planted. He asked for the hurricane lamp, took a heavy stick, and searched the garden, but he found no one there, and said that my mother must have been dreaming.' Mr Porter also recalled a two-storied half-timbered cottage at the back of his former home in Mill Street, which had a coach-house, stabling and loft. On one occasion the gardener had to search the loft because of mysterious noises, for which no explanation was found.

Bedford Corporation bought the Mill Street property from Mr. Porter in the early 1940s. It subsequently changed use to a British Restaurant, a local authority centre to provide cheap meals throughout the Second World War, before becoming the offices of the Housing Department. It was then variously used by other Corporation departments before ending up as the Probation Service's offices in 1966.

Paul Chapman, a local reporter, recorded an interesting meeting in 1972, with Derek Payne. Derek, then Deputy Clerk to Bedford Magistrates, had been working alone in his office late one winter's evening, towards Christmas 1962; his departmental offices were at thirty-eight Mill Street. Suddenly there came a sharp double knock on his door. 'I nearly shouted, "Come in", but I decided to go and open it,' he said. 'There was nobody there.' He was at the door in seconds but the hall outside was empty. He searched the building and found no one outside, and the entrance doors, both back and front, were locked. Was he sure it was a knock on the door? 'Yes, a sharp, distinct, double knock; like that.' He rapped twice on his desk. 'I'd heard people knocking on that door often enough; I knew that was what it was.' That was the only ghostly experience he had in the building. 'Looking back it seems a silly thing, but it made such an impression on me, I can still hear those knocks, they are so positive.' Did he think the knocks were the work of the supernatural? 'I've got nothing to back that theory; all I know is it was unexplained and it remains unexplained, I'm sure it was not a practical joke. There was a creaky wooden staircase and squeaky heavy doors which you had to slam to close. I'm sure I would have heard something if anyone had been in the building.' Derek's room was allegedly the most haunted in the building, it was situated on the first floor – 'a gloomy room in a generally dingy building,' he described it.

There were further reported incidents. One evening, around the end of 1966, at 7.30 p.m., Harry Larkworthy and John Harlow were chatting in the hallway when the front door handle began to turn. Harry was Senior Liaison Probation Officer for the Crown Court at the time. 'I remarked to Mr. Harlow something about wasn't it amazing that somebody had to turn up just as we were packing up for the night.' 'We were at the door in two strides, but there was no one there.' They looked all around the building; 'There was not a soul,' said Mr. Larkworthy. The ghost acquired a name – George, 'On account of his being a Georgian gentleman, or so we thought. We regarded him as a friend; he was almost part of the staff. I don't think anybody's ever worried about him, I don't think he has any evil intent.' Other probation staff confirmed the haunting. Miss Daisy Thurleigh recalled an experience with the Principal Probation Officer, Paul Hodgskin. 'We both heard a terrific banging. I thought someone had stumbled with a whole lot of files along the corridor, but there was nobody there. All the girls in the office below heard it and so did Mr. Hodgskin, but there was not a soul anywhere.' Interesting choice of words, because it may well have been a 'soul' that was responsible for these disturbances. Currently Mill Street is home to the Citizens Advice Bureau, and there seem to be no contemporary stories about any haunting, but then again the building would remain empty at night…wouldn't it?

One of Bedford's oldest and most infamous spectres is 'Black Tom', so called for his swarthy skin and greasy coal-black hair. He was a notorious highwayman, who struck such terror into the people of Bedfordshire that even after he died they took no chances with him. After he was hung on the gallows in Union Street, Bedford in 1607, a stake was driven through his heart to stop his spirit from walking abroad. He was then buried at the crossroads of Tavistock Street, Union Street and Clapham Road. A stake may be successful with a vampire, but 'Black Tom' was unaffected. Pretty soon the highwayman's ghost was being reported as haunting the vicinity of his execution place. A figure with a blackened face was seen staggering along Union Street, with head lolling to one side. As suddenly as he appeared he vanished, in broad daylight. The last witnesses to report seeing him did so in 1963, so perhaps we are safe now from his unwelcome attentions.

South Wing of Bedford Hospital has been mentioned to me on several occasions, by people who have worked there, as being haunted. I was fortunate in tracking down 'Mrs.Y', a friend of a friend, who was a nurse there in the 70s. One autumn night in1972 around nine o'clock 'Mrs.Y' was working, as a trainee nurse, on Russell Ward. It was a quiet night and the Staff Nurse and 'Mrs.Y' were getting the night medication ready. Suddenly a young girl came upstairs and into the corridor: she was quite pretty, with short brown hair and rosy cheeks, wearing a long dress and a long white coat; she was above average height and slim. The staff nurse was just unlocking the medicine trolley and checking the medicine chart as she came out of the office when she looked up she saw the young girl go into the toilet. She didn't come out again and the Staff Nurse asked 'Mrs.Y' to go in and check on her to see if everything was all right. The trainee could find no-one although she checked the toilets carefully; she did feel a distinct cold spot and found herself walking out of there backwards. When the two nurses compared notes they realised that the 'young girl' had not made any sound, but as the stairs were stone it would have been impossible not to make a noise ascending them. The staff nurse recalled that the figure had seemed to be 'gliding' along: there were no medical or surgical staffs on duty that night, so it could not have been a doctor. For the remainder of the shift the nurses didn't use that toilet instead they used the patients' toilets to empty their bedpans. If the doors to Russell Ward were locked at night they would often rattle as though someone was trying to get in but there was never anybody there. 'Mrs.Y' later learned that the ghost was that of a girl who had worked in the pathology laboratory; three years earlier she had committed suicide, when she was only twenty-three.

Shand Ward was a large long ward and one night, at three a.m., 'Mrs.Y' heard scuffling footsteps coming from bay five there. About half an hour later the odd shuffling began again, but no light came on. Both 'Mrs.Y' and the other Staff Nurse on duty heard the footsteps close to the corridor wall, come half way down from bay five to the nurses' station. It was odd, all the men in bay five were mobile but the steps sounded like those of a very old person, yet nothing was visible.

The night following a patient's death 'Mrs.Y' checked his empty room in the new block, right by the nurses' station, after rattling was heard; the man had been in the habit of rattling the raised sides of his cot to attract attention. Not only was the room empty but the sides of the cot were down. The nurse's mother too, had an alarming experience as a patient when she felt an invisible presence get into her bed while she was still inside, she spent the rest of the night in a chair. I am certain that there are many other stories about South Wing and North Wing is also reputed to have its ghosts; if anyone out there cares to contact me I shall be most interested to hear about them!

Ghost stories often appear more authentic when the sighting is made in broad daylight. The factors of fear and imagination are excluded, so we are more easily able to lend credence to 'daytime spirits'. Danny Ward allegedly had such a vision in the middle of Bedford, back in the 70s. He told reporters, 'I saw a monk with sandals on wearing a habit with the hood up so that you could not see his face; there was a chain round his neck and a large cross; he seemed to be looking down at the cross and meditating.' Mr Ward, a thirty year-old landscape gardener of King's Place in Bedford, saw the apparition walking past the Greyfriars public house. Other people in the street nearly brushed shoulders with the friar without seeming to notice him. 'I didn't think until later of the significance of the monk in grey, that's when I began to wonder if he was real.' Then he remembered that a few years previously a girl assistant at the nearby Carousel record shop told him that she, and others, had seen a phantom friar when working late; the shop site was later taken over by the Nationwide Building Society. A medieval skull was found in Priory Street in 1938 during the digging of an air-raid shelter and put on display in Bedford Museum. Mr Ward wondered, 'If disturbing bodies upsets the dead, who knows what being put on public display will do?'

Certainly, where now there is a concrete-and-glass landscape by the bus station, street names betray the earlier settings. Priory Street really did lead to a priory and real friars dressed in grey strolled through Greyfriars. All part of Bedford's rich and varied history which has seemingly left us with, among others, phantom nuns, World War One soldiers, highwaymen, Georgian gentlemen and monks – a veritable gallery of ghosts... Next time you are out walking in Bedford Town remember – that man or woman you pass in the street may not be as substantial as they appear!

Chapter 13

Mystery Mill

Driving through Hitchin onto the meandering B655 on a summer's day is a delightful experience.   My destination was the ancient village of Barton-le-Clay, Mid-Bedfordshire, which had a settlement as far back as the Iron Age; Barton's name means 'settlement where barley grows, and 'le Clay' was added to ensure no confusion arose with any other village called Barton (of which there are eight in England, plus a further fourteen 'Bartons' with additions)!   At the end of the large village I joined the A6, which no longer runs through Barton, considerably adding to its charm. A sweeping track just off the A6 takes you to The Waterside Mill, where a newspaper contact had put me on the trail of a ghost story. At the time of my visit the mill was a restaurant and tea rooms, but still recognizably a mill with an adjoining garden centre. My appointment there was with John Duggan, the genial white-haired Irishman who gave me a first-hand account of his experiences at this fascinating place. John had not long moved in as manager when, one winter's night after the restaurant was closed, he locked up and was reversing his car away from the front of the mill when he clearly saw a figure at the window beckoning him to come back. Concerned that he had inadvertently locked some poor unfortunate inside he rushed back, unlocked the door and frantically searched for the woman he had seen from his car. There was no sign of anyone even after a thorough search of the premises. Mr. Duggan was a self-confessed sceptic, but he admitted to a distinct shiver down his spine as he relocked the place and left for the night, convinced that what he had seen must have been a ghost. John was given a second opportunity a week later when again, as he prepared to leave, his headlights illuminated the apparently solid apparition at the window. His description: 'She was an old woman, probably in her eighties. What I particularly noticed was her long grey hair and bony hands as she beckoned me back to the mill.' He didn't return this time; instead he left in rather a hurry.

Some six weeks prior to my meeting with John a guest had been upset to hear the persistent crying of a child, while she was in the 'ladies'. No children were in the mill at this time.

'Psychic Phil', a friend of John's, was a local medium who detected a 'friendly presence' on visiting the mill, as did two ladies from a local psychic circle.   There had been other minor incidents, of things 'disappearing' minutes after being put in one place, only for them to turn up later in some other unlikely place.   Jo Williams, a waitress at the mill, attested to this as well as admitting to feeling decidedly uncomfortable in the upstairs function room. I ventured up alone (John didn't want to set foot up there) to the garner floor right at the top of the building. I climbed up some extremely steep wooden 'steps' (one was missing) where you could easily miss your footing – and I nearly did. I must admit it was a creepy place, dusty, with junk stored up there. I didn't linger and cautiously made my way down again.

The history of the mill at Barton-le-Clay goes back to the Domesday Book; the first hard evidence of a mill is in 1086. Barton, a village in the Flitt district owned by the Abbot of Ramsey, contained one small mill of taxable value two shillings (the lowest taxable value of any mill in Bedfordshire), Barton's population then being a mere 150. Over the centuries Barton has suffered mixed fortunes: in 1348 the Black Death visited the area at a time when it accounted for the killing off of about thirty per cent of Bedfordshire's entire population. Then in 1559, the year after Princess Elizabeth became Queen, a virulent strain of 'flu struck and over the next two years it caused the death of one in four Barton villagers. During the 1700s Edward Willes Bishop of Bath and Wells owned the Manor of Barton and he endowed a school in the village, which was eventually to become known as Barton Manor School. In 1872 hundreds of tons of dinosaur manure were dug up in the school grounds and when treated it was turned into valuable fertilizer. The sale of this funded a school extension and helped to ensure the continuance of teaching here until Barton Manor School finally closed in 1977. Lord Clarendon was to famously christen Barton-le-Clay the 'Switzerland of Bedfordshire', describing its delightful situation at the foot of the rolling hills, thought to be Bunyan's 'Delectable Mountains' in The Pilgrim's Progress.

A tragic well-documented tale of Barton concerns a traveller passing through one bitterly cold February day in 1874. George Buffam was a respectable man, a gilder by trade; he had walked the thirteen miles from Bedford. He tried all five inns but none of them would give him shelter; he made the same offer at several cottages but nobody would let him in, even though he was offering good money – one shilling for a bed for the night. 'Feel how cold I am,' he told people. 'Good God! I shall be left out to die in the street.' At 2a.m. next morning his body was found by the village policeman; he had indeed died of exposure in the street.

Meanwhile Barton's mill continued to change owners; it was unusual for villagers to have private ownership of land in the sixteenth and seventeenth centuries, but in 1611 the mill was owned by one Felix Wilson. Private milling came to an end in 1643 however, when William Burridge, the incumbent miller died. The owner Richard Norton, Lord of the Manor, outlawed all hand operated querns insisting that every villager had all their grinding done at the mill. By the mid eighteenth century the mill was operated by William Chesher whose family kept the place until 1831 when William Blott paid £950 for the mill, a house and pasture land; on Blott's death the mill was auctioned and sold to two gentlemen of the cloth – The Reverends William Gale and George Tuson. A most tragic story was about to unfold as William Hipgrave, a master miller, came in as tenant; when he died and his son George also died, it was down to the surviving son Fred Hipgrave, to carry on. He entered into an indenture in 1907 which placed the mill in his tenancy, complete with extremely severe conditions placed on its upkeep and maintenance. By the twentieth century things were going badly wrong at Barton Mill; as Luton Town increased in size large companies like Vauxhalls and Skefco drilled their own wells, to such an extent that water levels at the mill dropped alarmingly and the miller was obliged to work less and less. Fred Hipgrave suffered with chronic depression with some obvious causes; in addition to the pressures of his indenture a series of misfortunes plagued the mill. Severe winds blew the cart shed down, the mill wheel broke and the steam engine failed. It was becoming increasingly difficult to pay the rent; Fred missed his beloved father, master miller William. Fred's sister Mary and brother George had also passed away. He could not cope and was always asking for help when things failed at the mill. He seems to have had no skill at repairing things and he worried constantly. He had poor health; like his dead sister he too had 'the sugar' (sugar diabetes), which in those days was a fatal condition. It was not till 1921 that insulin came into use and made diabetes treatable. On Wednesday 3rd October 1912, Simeon Mead visited Barton Mill and discovered the body of his friend Fred Hipgrave, hanging from a beam in the mill house. Hipgrave had only been 54 years old. Two days later the coroner's inquest announced the verdict; suicide by hanging whilst temporarily of unsound mind.

In 1913 Mr Hynd, a schoolmaster from Hexton, bought the mill; his son Harry used a Luton firm, Guerneys, to restore it to good working order, but in 1915 Harry was called up for armed service. Albert Hodge was the last miller at Barton; he only used the mill part time to grind animal feed for his cattle. In 1924 a storm caused the mill bank waters to overflow and flood the mill; this was the end of work at the mill which remained empty from 1926 onwards. A Mr L.W. Vass restored the mill in 1974 and it became a listed building, but sadly, by 1998, after Mr.Vass's death Barton Mill became quite derelict again with almost every window smashed and large parts of the weather-boarding removed or damaged; attempts had also been made by vandals to burn the place.

Now the Water Mill had a new lease of life as a busy restaurant and tea rooms, and John Duggan was a personable host. I was able to nail one myth for him; local legend would have it that Fred Hipgrave hanged himself and his 'wife' died three weeks later 'of a broken heart'; my researches revealed that Fred Hipgrave never married! John told me he had heard footsteps upstairs when the place was empty and as nips of whisky seemed to disappear from his bottle of Scotch he nicknamed his ghost 'Casper', because it seemed a friendly if spirit-loving spectre! I took some photographs while interviewing John; the first one I wanted to take was of him posed at the window where the ghost had been. My trusty Olympus compact wouldn't work – twice; undeterred I tried a third time and at last the shutter released, yet when I went outside to photograph the exterior of this picturesque spot my camera was working perfectly and has never given any trouble since. There is something odd about the mill.....but is it haunted, and if so by whom?

John Duggan left Barton-le-Clay and the restaurant and tea rooms he used to manage are no more. Instead the old mill house has been converted into a busy arts and crafts centre, but maybe the ghosts still prowl there at night...

I was recently in correspondence (October '04) with the prolific author, and ghost hunter supreme, Peter Underwood, who I have already mentioned. He recalled investigating a haunted, derelict mill near Hitchin where there had been a suicide. As it was one of his earliest cases and it was over 60 years ago he couldn't be certain, but it seems highly probable that it was the same place – Waterside Mill.

Chapter 14

Village Visitations

The ghosts of Bedfordshire are not confined to the bustling County Town of Bedford or even to smaller market towns like Ampthill. Often they appear in more rural surroundings, as the following stories will show.

Kensworth's Haunted Footpath

Kensworth's Footpath Number Four is also known as the Coffin Route, from Common Road to Hollicks Lane. If you want to avoid seeing an apparition you may be advised to avoid this route as it has gained the reputation over many years of being haunted.   Albert Sipson of Dovehouse Lane Kensworth researched the history of the footpath. He discovered that it became known as the Coffin Route because in times past people who could not afford transport for a funeral would carry the body along the pathway to be buried in St. Mary's Churchyard.   Three ghosts are alleged to haunt the pathway; a witch, a headless milkmaid and Shuck, a one-eyed large black Retriever. Old Shuck is well known in Norfolk and other parts of East Anglia. The name is derived from the Anglo-Saxon scucca or sceocca, meaning Satan.   As for the other strange visitors I can offer no explanation.

Kempston's Haunted Boathouse

Nowadays Kempston is virtually a suburb of Bedford Town, but not so long ago it was a rural area where sheep grazed in tranquil meadows. The story persists of the unfortunate shepherd whose faithful Alsatian suddenly turned against him one day.   The unsuspecting man was savaged to death by his dog. On the site where the shepherd used to lay under the stars there is now a boathouse, off Hillgrounds Road which belongs to Kempston Youth Club/Outdoor Activity Centre. This area also contains the ruins of old Kempston Manor where King Henry III stayed in 1224 and Devorgilla (mother of John Balliol, King of Scotland), lived till her death in 1290. As with many other ancient sites, tales of the supernatural abound here. Clem Tite was a storeman/technician there until leaving in 1990. He commented, 'Almost every afternoon in the summer I used to hear footsteps in the building. I used to search the place, but there was never anybody there.' Other people also heard footsteps: on one occasion one of the youth workers brought her dog in. She went down the passage to a room at the end of the corridor where she was hit by an 'absolutely freezing' sensation as if she had walked into some giant freezer. It spooked the dog, which howled, while its hair stood on end and it ran off with its tail between its legs. The woman involved was Jackie Jennison – her dog was an Alsatian. 'My dog Evie just wouldn't go into the room', she said. 'One of our cleaners also had a dog and that wouldn't go in there either.' Did the ghost of a long-dead shepherd stalk the boathouse and grounds, warding off any dogs who venture too near?

Riseley's Haunted Farmhouse and Other Psychic Phenomena

Riseley is well-known amongst clay pigeon shooting folk for the excellent facilities available at Sporting Targets, particularly the challenging high towers and well laid out sporting layout and skeet ranges. This top flight shooting ground has been owned and run for many years by the genial Ken Surridge. When he was bringing up his young family in the 1950s Ken lived at Buryfields Farm. About halfway through the village there is a turning into Keysoe Road, leading off up the hill to the right. This is the way to the most haunted place in the district, Buryfields Farm. Ken remembered some chilling incidents: lamps would light up for no outward reason, and once a pump with a lead-based solid structure collapsed, striking and hurting his daughter Christine. She and her brother Allan would wake up at night terrified, and on one occasion Christine saw a door and a dark figure in a place where there was ordinarily just a solid wall. The most sinister aspect of life at the farmhouse was the behaviour of the outside doors; even when securely locked they would open at night. In a desperate attempt to secure his home Ken drove six-inch nails into a door to make it fast one night, but by morning, with nails intact, it was open once more.

The family dog Spider hated going into the house. When mediums arrived to try and 'lay' the ghost Ken had to carry the reluctant dog inside, holding it tightly. Ken said that at Buryfields it was more of a feeling, a sense of presence, than anything really tangible.   He recollected one instance when the stockman, who was in a nearby field, heard screams coming from the servants' quarters.   On going to investigate, the stockman found a terrified girl in one of the rooms; according to her there was a menacing man present, but the stockman could see nothing. Ken's most frightening experience however, was not at the farm at all: one day he was on the road out of Riseley on the way to Dean when he had a weird encounter with the supernatural at Eastfields Corner. Something sounding like a horse crashed through the hedge and crossed the road from the fields on one side to the other, but nothing whatsoever was visible. Ken and two other witnesses present at the time could hear the breathing and feel the rush of cold air as it passed.

By the side of the road that runs from Bletsoe to Riseley is Tom Knocker's Pond. This spot also has the reputation for being haunted, by the spirit of someone who drowned there long ago. Further on in that direction people have encountered the Tidbury Ghost. Eric Hancock, a local resident, encountered a white figure there late one night. It crossed the road from the left and then vanished as it started to walk over the fields towards Tidbury Farm. No sightings have been reported in recent times however, since the farm was swallowed up by the Royal Aircraft Establishment.   Rumours persist that there is a path within the R.A.E. grounds which made the ministry police dogs bristle.

Further along the village street is the Fox and Hounds, long reputed to be haunted by a former nurse killed in the last century. It appears that she slipped and fell under the hooves of coach-horses and was trampled to death. In those times casualties were taken to the nearest inn, which happened to be the Fox and Hounds. She favours haunting near the fireplace in the lounge bar, but has also been seen crossing the car park. Her favourite target is the lights, which she switches on during the night. On one occasion she even managed to turn on a cupboard light that had not worked for months and refused to function again the following morning.   Reports are mainly confined to audible phenomena these days, footsteps and other noises overhead, plus the occasional ghostly cough when there has been nobody (living) upstairs.

On higher ground away to the left is the village church. In the late 70s a local family had an eerie experience when walking down Church Lane late one night. As they looked towards the church they could see straight through to the altar and the church walls appeared to be in ruins. This was just one more example of the varied psychic phenomena that make Riseley a candidate for 'Bedfordshire's Most Haunted Village'.

Millbrook's 'Magic' and 'The Ghosts Who Never Were'

Millbrook is perhaps best known these days for its outstanding car proving ground, owned by General Motors, and often visited by both press and television. The test facilities are second to none. This attractive village also boasts a fine golf course and gets my vote for probably the prettiest village in the county. Its setting is secluded, just off the Ampthill to Woburn Road, in well-wooded terrain. It is a place steeped in legends and one of the few places able to challenge Riseley for the 'Most Haunted Village' title. One theory about Millbrook is that it was the inspiration for John Bunyan's 'Valley of the Shadow of Death' in The Pilgrim's Progress. A particularly haunted area was at the junction of a path from Woburn Road to Millbrook Church. Long ago travellers returning from the Millbrook Feast would pass this way; the pub on the Woburn Road where the Feast was held became a farmhouse. The junction had many uncanny tales, one involving a woman and child who passed a pile of stones at night and saw them rise up and whirl around them. Another woman at the same place saw a large black object come over a ten foot hedge and land in the road without a sound. A third incident was reported by a man who said that he was accompanied by a strange light which appeared at the same hedge and later disappeared. Yet another man claimed that he put his foot on a stile near the spot and suddenly found himself in Clophill.

'Galloping Dick' was a highwayman who preyed on stage-coach travellers on the Woburn Road, a main route for coaches bound for Oxford or Cambridge. He lived in a tumbledown dwelling by the sand-pit on Millbrook Hill. He was eventually caught and hanged for his crimes but his ghostly presence continued to be felt in the village where he lived; late at night the pounding of horses' hooves was heard, but nothing was ever seen. These visitations were attributed to Dick. A headless horseman haunted the winding road that leads to Millbrook's tiny railway station, allegedly another highwayman, but I see no reason why both hauntings were not attributed to 'Galloping Dick'.

Another legend which persists in Millbrook village is the 19th century tale of the unfortunate young woman who encountered an apparition of a black dog. To meet one of these spectral hell-hounds was an omen of death in the family. They were a fearsome sight, 'as big as a calf', with red, glowing eyes.   The girl was dead three days later. Interestingly there are many stories about these canine spirits, sometimes called 'galleytrots' that haunt the lanes of East Anglia. The unusual name is thought to be a corruption of gardez le tresor or 'guard the treasure'. If legend is to be believed galleytrots are mostly found near old burial grounds or hidden treasure. This could fit in with another Millbrook story, that there is a pot of gold buried in Moneypot Hill. Black dogs were also associated with Diana, the witch goddess, whose cult was extremely strong in country areas. I am often asked why black dogs in particular should be associated with a forthcoming death. I subscribe to the view that some level of the mind already knows about the future; this is its method of conveying the information symbolically.

Millbrook's most interesting Victorian 'ghost story' concerns the Huett family. Once there was a grand Tudor home and an elegant altar tomb to commemorate them. The home has long since gone and the tomb which stood by the church chancel with full length effigies of the Huetts also went. During restoration work in 1857 the tomb was removed due to its dilapidated state. The two figures of the Tudor knight William and his lady Mary Huett, both carved in grey stone remained in the church. Soon weird cracking noises (said to sound like whips) and groans were heard in the church. Superstitious villagers concluded that the Huetts were showing their displeasure at not being allowed to rest in peace. A hasty decision was taken to remove the Huetts to the rectory cellar but the 'protest' noises increased in both loudness and frequency. Finally the effigies were buried, with full exorcism rites, in consecrated ground in the churchyard. Still the weird noises continued, until 1888 when the chancel roof collapsed and the natural explanation for the noises became apparent. The roof was replaced and the noises were never heard again, thereby ending the story of the 'ghosts who never were'. In 1919 the Huett effigies were excavated by boys of Bedford Modern School Archaeological Society under the guidance of the school chaplain and the Rector of Millbrook, and now they lie peacefully, side by side in Millbrook Church.

At various times a spectral glow has also been recorded, emanating from the windows of Millbrook Church when it has been empty. The writer Menzies Jack summed up the scene particularly well when describing the church, he called it 'this little bit of heaven on a hill.'

Silsoe Sarah

In 1987 Mrs. 'Mabs' Manby was interviewed by a reporter from the Luton News about her lovely old home in Ampthill Road, close by The George Hotel in Silsoe village, just south of Clophill. Mrs. Manby's house was believed to have once belong to the de Grey family's estate agents (the de Greys owned the Wrest Park estate, now in the care of English Heritage). If all this sounds familiar it should do: over a hundred years previously this was the site of pioneer ghost hunter Bill Turner's first ever ghost sighting (see the chapter 'Pioneer Ghost Hunter'). 'Mabs' explained how she had seen a small apparition many times in the twenty five years that she and her husband, David had lived at the house. 'I do often see it cross the hall it's a small white figure,' she said. 'One of my friends was here one day and she told me: 'Something has just flitted across your hall,' Mrs. Manby claimed to see the ghost about two or three times a year, and believed it was the spirit of a small girl. Mrs. Manby was unconcerned about her little 'guest'. 'It's not at all alarming; it's something I just live with.'

It seems clear that Bill Turner's ghost-child 'Sarah' continued to search in vain for her beloved nurse nearly sixty years after Bill's death. Over the intervening years her haunting gradually diminished in intensity until she confined herself to the briefest of appearances without the creaking floorboards, door-knocking and hand-touching that Bill had experienced during Mrs Hallam's residency a hundred years previously...

The Marston Moretaine Exorcism

In 1986 the Reverend John Greenway, rector of the Anglican Church in the village of Marston Moretaine, was called to a home in Hillson Close. Janice Green had called him in after suffering with paranormal phenomena for fifteen years. It began with things disappearing in the house and Janice would often wake suddenly at night and be drawn to the window to look at the bottom of the garden. The vicar conducted an exorcism to banish what he considered an evil spirit that lingered in the lady's garden. Janice had also been bothered by poltergeists, possibly connected with previous occupants of the house, but after the exorcism these manifestations were stopped as well. Janice discovered that her garden was on the site of an ancient pond which could have been used in past centuries as a ducking-pond by witchfinders.

Lost Lamb of Arlesey

I picked up an interesting story while talking with Mrs. B., who had lived for several years at number 28 Lamb Meadow in Arlesey. She was repeatedly visited by the most vivid and disturbing dreams. They would happen in the hypnopompic state i.e. that state immediately before waking up. This was back in 1984. A perfectly clear vision of a little girl would come to Mrs. B. Even now nearly twenty years on she could recapture the image, it was that haunting. The child was about nine years old, of Italian appearance. She always had the same plaintive message, 'Can I have my daddy back?' Mrs. B. asked her family to say prayers and the 'visitations' abruptly ceased. There was however, a continuing 'presence' that the lady felt in her house and she was certain it was a man. A visiting female friend also strongly sensed this 'presence'; it was as though 'someone' was constantly sitting by the stairs, it was also sensed in the kitchen. 'He' sometimes brushed against people. On one occasion a securely fastened bottle of bleach had unaccountably leaked, out of the cupboard and onto the carpet. The house was built on what older residents remembered was a chicken farm and allotments. Did the spirit of a dead father wait there to be reunited with his little daughter?

Bolnhurst's 'Cedric'

Ye Olde Plough, at Bolnhurst is a delightful long, thatched old building, set well back from the road in its own sizeable grounds. It hadn't been used as a public house for nearly four years when I met Rita Horridge, who had lived and worked here for over twenty years and she ran the 'free house' pub from 1983 to 1999 with her 'ex', Mick. She was forced to close the pub in March 1999 as it was making a loss of £2,000 per month. Now Rita has left the area, but she used to run the Plough as a 'Spiritual Sanctuary' where she would organize faith-healing, monthly clairvoyant evenings and 'The Sunshine Group', which had talks on various spiritual topics, also discussions and meditation for those with an interest in improving their lives. Some of the 'Sunshine Group's most interesting workshops included Crystal Clairvoyance (which I have attended), Growth of a Medium and Sound Healing.

The history of Ye Olde Plough dates back to around 1480 when it was the house of a farm called Brayes, on Edward, Duke of Buckingham's land. During the Great Plague in 1665 the villages of Colmworth and Bolnhurst were struck, probably through the carriage of clothes donated by rich Londoners to the poor of the county. The only solution to the pestilence was fire; accordingly the entire village was burnt to the ground leaving just the church (as was reputed to be the case at Clophill's St. Mary's church) and Brayes, spared as they were deemed to be on the edge of the contaminated area.

Ye Olde Plough had a ghost called Cedric, described by Maureen Bishop, a Stevenage medium, as wearing a white shirt, with rolled up sleeves and a waistcoat tied at the back. He was a farm worker who liked living here when it was a farmhouse; he had a large stomach kept in by a big leather belt and was about six feet tall. Rita's friend Tricia also practised as a medium but she used crystals as her focus and she too received mental pictures of Cedric many times; he wore big boots, his trousers were tied around with string, and he was a benign presence.

Another friend, June Sturgess, stayed the night at The Plough and felt a dip as her mattress went down, as though someone had sat on the bed, which caused her to tell 'Cedric' to go away as he was frightening her. In more recent times the ghost's presence has been felt rather than seen or heard but he remains 'resident' at the place where he was supposedly happiest during his lifetime.

The Gamlingay Prophesies

In the eighteenth century there was a tavern called The Bull at Church End Gamlingay, which closed down and eventually became a derelict cottage – 19 Church End. In 1969 Mike Harris bought the property and set about renovating it. 'After I'd been working on the cottage for three months, I began to notice a shadow going past the open doorway,' said Mike. 'In the end I said if you want to come in and talk then come in and talk.' The ghost, who called herself Abigail, replied, "I don't know why you're working, your wife will never come here." 'Three months before I finished the work my wife left me, so Abigail's prophesy had come true. Abigail told me she had come to the tavern in 1720 with a baby but she was unmarried. Her lover and the father of the child was 'a gentleman highwayman'. When the child, called Marcus, was eleven his father came and took him away to give him a better home than Abigail could give the boy, as she was only a servant. She was hysterical when they took the boy away.'

Mike moved into the cottage in 1973 and soon after Abigail made a second prophecy: 'She kept telling me I would meet someone else who would be blind in one eye and there would be a son of that person I would meet and the name would begin with a 'G'. As my wife had left I decided to advertise for someone to help remake a home and I had fifty-four replies. The second was a woman called Gillian who had a son called Marc and she was blind in one eye. As soon as I met Gillian I avoided her because the prophecy was coming true. But after six months I married her and took on her son Marc. It was like history repeating itself,' said Mike. 'Marc told us the lady came to him at night time and said she was going to look after him,' said Gillian Harris. 'We went to a spiritualist and we took a thimble we had found which had belonged to Abigail and he confirmed all Abigail had told us,' said Mike.

He added, 'The spiritualist would never come to the house because he said Abigail's presence was too overwhelming.'   In 1978 the Harris family moved out of Gamlingay to farm a smallholding in the Fen country – near Wisbech. Abigail moved with them, according to Gillian Harris's statement in 1979.   'When we came over to Wisbech in the van Abigail came with us and she's here now but she is much quieter than she used to be.'

These 'Village Visitations' neatly illustrate an answer to the question – where are you likely to see a ghost? Our stories have encompassed private houses, a pub, farmhouse, a footpath, country lanes, even a boathouse – so the answer to our question has to be – you can see a ghost anywhere, and Bedfordshire is a good place to start!

Chapter 15

Priory Phantoms

In any psychic poll of Bedfordshire Chicksands Priory would have to rate in the top three supernatural 'hot spots' for the county. What make the hauntings here stand out are the volume, regularity and longevity of the ghostly activities experienced by people not given to flights of fancy. Indeed many witnesses have been trained observers, disciplined, well grounded servicemen of both the RAF and the USAF. The principal resident phantom is supposedly Berta Rosata, according to the legends surrounding Chicksands Priory. She was the classic 'fallen nun' and her crime was to fall in love with and become pregnant by a canon of the same society, the Gilbertian Order. Rosata's punishment was to be walled up alive, but before the last few bricks sealed her fate, she was forced to watch her lover's beheading. There is a plaque in the cloisters that bears the script:

Moribus Ornata Jacet
Hic Bona Berta Rosata

This translates as: 'By virtues guarded and by manners graced. Here alas is fair Rosata placed'. Rosata's ghost, the legend continues, haunts the Priory in search of her dead love, on the 17th day of the month.

So much for myth and legend now for some facts: the plaque commemorating Rosata isn't on a cloister wall at all, as in monastic times it would have been the inside wall of a cellar. Rosata was definitely not a medieval name but seems to be an invention of a much later age, possibly 18th century, to lend a romantic air to Chicksands. What is known for sure is that a manor was granted to the order in 1150 by Payne de Beauchamp and his wife the Countess Rohese of Bedford Castle; not far removed from the name Rosata. The Julian calendar was in use during monastic times; the Gregorian calendar superseded it in England and Wales in 1752, so the modification would mean that the haunting date, instead of being on the 17th of every month, would now be on the 28th of the month.

Descriptions of phantoms seen at the Priory vary enormously, leading me to assume that the site is visited by a number of ghosts; indeed nine uneasy spirits are reputed to be associated with the building. Hauntings are widely dispersed too. The King James Room and Pigeon Gallery are the most frequently mentioned although a number of sightings have been witnessed in the grounds.

First recorded in the Domesday Survey of 1086, Cudessand Manor was owned by Hugh de Beauchamp. His grandson then granted the manor to the Gilbertians. Unlike most orders, which were founded abroad, the Gilbertians were founded in England, by St Gilbert of Semperingham, a native of Lincolnshire, with their priory at Cudessand (as Chicksands was known then). They were one of only nine religious orders in England to enclose both men and women. The priory grew to eventually house fifty-five canons and a hundred and twenty nuns, who were supposed to be strictly segregated. Shortly after the Priory was established St Thomas a' Becket is said to have sought succour here whilst in internal exile from Henry II. Gilbert himself is known to have visited the Priory several times. In 1535 Dr Richard Layton reported to his master Thomas Cromwell that he had found two of the nuns 'not baron', one had been made pregnant by a serving man, the other by a superior. Whether the report was true or not and there is some doubt, the nuns' punishment would not have been as savage as the old legend portrayed. Thomas Cromwell would have enthusiastically seized on any improprieties among the religious orders to report directly back to his King, Henry VIII. Not for nothing was Thomas called 'malleus monachorum' (hammer of the monks). The Dissolution of the Monasteries followed, beginning in 1536 by which time the Priory had suffered many setbacks. The Prior and Prioress surrendered in 1538, with just six canons and seventeen nuns left to be pensioned off. Even a King as powerful as Henry VIII would not have had the authority to break the monasteries alone, had they not grown corrupt and lost touch with the people. If they had remained popular the Dissolution might not have occurred.

Chicksands, now Crown property, was leased to Thomas Wyndham and later conveyed to the Osborne family (the 'e' on the end of their name was dropped in time). It was to stay in their hands from 1553 onwards for the next 400 years. Sir Algernon Osborn sold it to the Commissioners for Crown Lands in 1936. In 1939 the Crown leased Chicksands to the Air Ministry. It became a Royal Air Force Signals Station throughout the Second World War.

At the end of wartime the Priory was variously used as an officers' mess, an airmen's quarters and a bachelor officers' quarters. In 1950 the Tenth Radio Squadron (Mobile) of the United States American Air Force moved in; they eventually vacated the building in 1970 as it failed modern fire safety requirements, but they continued to use the many other modern military buildings on the extensive site. The ghosts had the Priory building to themselves for a considerable amount of time. US Forces left their Chicksands base altogether in 1995 and following extensive renovations to the Priory in 1997 the Defence Intelligence and Security Centre (established in 1996, by the RAF, Army and Navy, to oversee covert intelligence operations) and the Central Officers Mess to the Intelligence Corps moved in. So now there are two kinds of spook at Chicksands!

That is a brief history of the last 850 years of the Priory, now for a history of the hauntings. There are two cases which date from the First World War; the first involved a maid, Annie Stamp, who had worked at the Priory for 30 years. She was taking a glass of hot milk to the King James Room at about ten in the evening of a winter's day in dim light. 'A tall fascinating woman dressed in white' flashed past her in the picture gallery. 'I had not been drinking,' said the witness. 'I dropped my tray and fled in terror, I am not sure it was Rosata, but I'm sure it was something.' Stranger still was the case of the man in the same period, World War One, found dead just outside the Priory. His black hair had literally turned white overnight and his features were frozen into a look of terror, so it was said that he had been scared to death.

During the Second World War an RAF officer's practical joke misfired. Dressing himself up in an old sheet he stood outside the King James Room; his intention was to give one of the girls a fright. A young lady duly turned up and gave a shriek of fear, whereupon the officer returned to his room to change. A third party, upon hearing the scream, came to investigate – and bumped into a real ghost! It was a woman in black with dishevelled hair partly obscuring her old wrinkled face. The revenant then disappeared – straight through the wall.

The 1950s were a busy time for hauntings. George Inskip had been Head Gardener at the Priory for 30 years. One day he was in his greenhouse tending to the grapevines, but as he looked up he saw 'a dark, greyish shape' coming up the pathway. When he went outside however, there was no-one there. 'It was certainly not a shadow,' he said. 'There has always been ghost talk but I never believed it, now I'm not so sure. Whatever I saw was wearing a cowl and came straight up the path; it was dark, and if I had been outside I could have seen it more clearly, but I was looking through the glass of the greenhouse. If it was a ghost I don't know, but I'll never forget it.' Subsequent excavations in this area revealed a number of graves of long-deceased nuns and canons...

In 1954 an RAF officer reported seeing 'a ruddy faced woman with untidy hair, holding a notepad, wearing a dark dress with lace collar.' She was standing at the foot of his bed. Three years later an Air Force Lieutenant was sleeping in the officers' quarters when he was violently clawed on his left side and dragged in that direction by a ghost. 'Imagine my surprise,' he was later quoted, 'when I saw not a savage apparition but an intensely illuminated youthful face, smiling at me in the most friendly and intimate manner.' Immediately after his sighting the apparition shot away at terrific speed, receding into a pinpoint of light and disappearing with a hollow ringing cry. These and other authenticated sightings were gathered together by three young American airmen stationed at Chicksands. They overcame many obstacles to collate, edit and print a private book of these phenomena; they were Richard Humphrey, Paul Lajoie and August Trottman.

About ten years ago reporters from the Bedford Herald & Post staged an all-night vigil at the Priory. They based themselves in the King James Room with plenty of coffee, sleeping bags, torches and cameras. They patrolled the corridors and twisting passages, determined to confront any ghosts that might appear. Although no apparitions materialised they were not entirely disappointed. Several inexplicable events occurred; the batteries in their camera and freshly-charged flash-pack all mysteriously drained down. A door in the empty building slammed (it seems there was a history of this regular occurrence) and they also witnessed an exterior security light extinguish itself while they watched. Most convincingly of all, one of the photographs taken during the night had a most strange image on it – despite being taken in an empty room. It was an odd streak of light that just could be a cowled figure...

Defence Intelligence and Security Centre currently occupy Chicksands Priory, which continues to produce manifestations. Fairly recent experiences include balls of white light seen floating in mid-air, as seen by a reliable witness in the billiard room. Another member of staff reported seeing a figure thought to have been the phantom nun of legend. Other witnesses have heard children's laughter and seen lights emanating from empty rooms. Two guard dogs with perfect obedience and bravery records have had to be sent for retraining after refusing to go anywhere near the Priory. Certainly Chicksands is the sort of place that holds a strange fascination for some people, like Mr Roger Ward of Caldecote. He has compiled a pamphlet on the history of the Priory, and also works as a volunteer, restoring this unique building. Mr Ward first became involved in 1973, because when tracing his family tree he discovered that he was related to the Osborn family; Sir Danvers Osborn has become a personal friend. Mr Ward's involvement has grown so much that he now spends all his spare time at the Priory. His ambition is to get the building restored, spend the rest of his life there and be buried with the nuns and monks at the front of the building! He gives guided tours along with other Friends of the Priory every third Sunday of the month; there is much to see with sixty-six rooms, including the chapel. Such historic and haunted buildings are few and far between.

Nearby Chicksands Wood is also reputedly haunted: a report was received in 1971 from a startled motorist who suffered a most unnerving experience. While driving through Chicksands Wood he came face to face with a man on a white horse that rode out of the woods and galloped straight through the car!

One of the most persistent stories about Chicksands Priory concerns an airforceman who was cycling by the Priory when he saw an apparition. Shaken by his unexpected encounter he told his wife, who accused him of being influenced by alcohol! The following February he was close by the Priory again when he had the same unwelcome meeting with something that looked like a nun. The spirit came from the river and ascended near the south-east corner of the building, but as it reached the second floor it disappeared. Local historians agreed that, in the eighteenth century, there was an outside staircase because it showed up in the 1730 print of the Priory. The airforceman would have known nothing about that.

It is unfortunate for ghost-hunters that Chicksands is now a part of the Defence Intelligence and Security Centre, which makes it a most sensitive building indeed, and off-limits to those of us outside the intelligence services. Past vigils have almost always produced paranormal phenomena and I see no reason why that should not be the case now. This 850 year old abbey must have many memories, echoes of the past that may be recalled today. I did write to D.I.S.C on the subject of holding an all-night vigil and received a prompt and polite reply. I was thanked for my letter, but as Chicksands is a fully functional military base and the Priory is home to many officers on site, my proposed vigil would not be possible.

After the first edition of this book was published I was contacted by Roger Ward, who had read it and enjoyed it, particularly this chapter about Chicksands. He was able to add some interesting observations, which was gratifying because he had started out as 'a total sceptic' to use his own words. Roger has lived in Bedfordshire all his life and he has three passions – music, history and religious studies. He is a piano-tuner by profession and he was introduced to Chicksands in 1966 when he was engaged as the organist for the Protestant Chaplain, on a six month contract to the United States Air Force. He was to stay on for over thirty years… Since 1973, while tracing his ancestors, the Priory and its history has became his 'baby'. He knows more about the site than any man living, having spent thirty years researching the place but he still manages to keep finding out more; with the arrival of the internet new facts keep emerging from people who have been in touch from all around the world, research can be addictive!

The story of Chicksands Priory was largely one of neglect for two decades; from 1940 to 1962 it was virtually untouched. When the USAF moved out in 1971 the place remained empty for many years so it was eventually opened to the public by the Friends of the Priory (whose Patron was Sir Stanley Odell, Roger was made Chairman in recent years) in order to raise money to restore the building. It was estimated that 100,000 visitors per year would come to see Chicksands if it was opened up like a Stately Home. In the meanwhile the Ministry of Defence had kept it wind and weather-proof. This beautiful old building possesses 66 rooms, 4 bathrooms and 3 toilets, along with 13th century roof beams and plenty of ancient stained glass. The last tour was in 1996 and then the Priory was closed for total restoration at a cost of over £5,000,000. The public was readmitted in 2004 and the Sunday tours continue to be a great success.

Roger has had several brushes with the ghosts of Chicksands Priory. In 1999 a local ghost hunting group brought along the clairvoyant Marion Goodfellow, who has been a guest on Living TV's Most Haunted. Roger arranged to briefly show the group around but ended up staying till 3 am as he found it all so intriguing. When he checked Marion's findings they were exactly in accord with those of a psychic medium who had visited back in 1975. Roger said 'It was as if the first clairvoyant had returned after 25 years.' Another medium, visiting in 1977, had described paintings and furnishings in some of the rooms in the old building and many years later Roger came across an 1893 photograph showing the décor exactly as the lady had described it.

There are no written accounts of hauntings at the Priory till the start of the twentieth century, in 1905/6 the first references to the ghostly nun appear. On three occasions Roger has spent the night in the King James Room – it is the most haunted room in the house. The radiator has gone stone cold in a couple of instances during vigils, yet next day it has been found to be working perfectly. He didn't see anything but on one occasion he heard Gregorian chant being played on a phantom harpsichord – he later discovered that one of the Osborne family (who owned Chicksands) was an accomplished harpsichord player...

Roger is convinced that there is more than one ghost haunting his beloved Chicksands, 'or they have a very large wardrobe.' He cited a case where a couple who lived in one of the base houses which had been built in the 60s, had invited him to tea and told him that they had 'funny experiences'. Their dog kept barking at an unseen presence and things had unaccountably moved about in the house, enough to convince them that their place was haunted. Within a year they had moved and a new family was complaining about the house. Eventually a new Chaplain moved in and he called in the Bishop of Bedford, who visited one Sunday to bless the building, which is still standing on a part of the estate where the carriage drive used to wend its way through the grounds.

I finally got to see the Priory for myself on Sunday 4th September '04 with about 20 other members of the Anglia Paranormal Investigation Society and we were amazed at how plush and new the interior appeared. Roger Ward himself showed us around on this fascinating tour. Roger's enthusiasm, wealth of knowledge and sheer love for Chicksands is immediately apparent to everybody and I would thoroughly recommend visiting Chicksands for a special day out. It was particularly interesting for me to visit the Pigeon Gallery (so called because of the decorative pigeons that were painted on items in Sir George Osborne's extensive collection) for this was the area where some quite spectacular paranormal effects were witnessed by investigators during a series of all-night vigils back in 1994/95. Currently the Ministry of Defence holds joint copyright on the reports (with David Wilson who led the investigations), which are quite amazing documents. One day I will get their agreement to publish details of these unique experiences but that is for a future book...

Roger showed us where Annie Stamp saw the ghost of a woman crossing the Pigeon Gallery, and there are reports of a nun also being seen here, she appeared to walk through a solid wall but it was discovered that a door had existed on the spot where the nun disappeared. Our guide then led us into the King James Room, so named as the block was built to contain King James's State Bed, bequeathed to the Osborn family before 1789. The bed is now at Kensington palace. The haunted room was built in 1806 for General Sir George Osborn, over the presumed site of the Priory Church. An impressive room, it is in the style of the Chapterhouse at Peterborough, an airy, lofty place with a fine vaulted ceiling. My eye was drawn to the huge window which ran from the ceiling almost to the floor, curiously in the stained glass at the top there was a Latin inscription, 'speculum verissimum' which translates as 'consider the truth', together with the rather spooky image of a skull. On the deep blue painted walls I was also attracted to the portrait of Vice Admiral (in Henry VIII's Navy) Thomas Wyndham, who had leased the Priory from 1540-1547. Thomas was a formidable, rough-looking man, with a crooked nose; I could well believe the rumour that he had been a pirate.

We were also shown up into the attics, which were a scary place to be ten years ago during the vigils, when there had not even been electricity up here. At least one extremely frightening event was recorded in the long corridor; it was easy to imagine how atmospheric it must have been in those days. Now it is all fully carpeted and freshly decorated, all the rooms leading off the corridor contain comfortable accommodation. It was interesting to see the oak and chestnut roof timbers through a small door at the end of the corridor, some of these timbers have been dated to 1468. After the tour I had a chat with Roger over tea and biscuits, Chicksands had worked its magic on me and I became a member of the Friends of Chicksands Priory. If you want to see the Priory for yourself tours can be arranged through Julie Benson, telephone 01525 860497 or Email: juliebnson@aol.com

Not surprisingly perhaps, Chicksands stands on a ley line, and of the 69 parish churches in North Bedfordshire all of the older ones are sited on these 'energy lines', which are often associated with hauntings and other paranormal activity.

Chapter 16

Luton's Lares and Lemures

Founded in the sixth century, the place we now know as Luton was originally called 'Loitone'. The first farm or settlement was called a 'tun' and was situated on the River Lea. Lea is thought to be Saxon in origin and meant bright, hence 'settlement on the bright river'. By the tenth century Luton had developed into a market town and its population had grown to several hundreds; most of them farmers, for the town's market would have provided a focal point for surrounding settlements and villages. By the next century (the Domesday Book of 1086 holds a record) Luton's population had risen to 750-800, which would have been a sizeable town since the average population for a village then was 100-150. With the arrival of the Middle Ages (from AD 1,000 to 1,453) the population had expanded considerably – to some 1,500 souls.

Ghosts have been with us as long as recorded history. The Romans called their 'good' ghosts lares (lares domestici were household ghosts, like poltergeists). Lemures on the other hand were 'evil' ghosts, who often tormented relatives. Maras (from which we get nightmares) were evil tormenters that plagued people at night. Luton has its fair share of history, ghosts and strangely atmospheric sites.

Galley Hill on the outskirts of the town is one such area with ghoulish associations, as it was once used for public executions. A gallows stood here and the site was also a burial ground for local witches hanged during the persecutions of the 16th and 17th centuries. Many an innocent old wise woman has gone to her death denounced by someone with a grudge against her. During excavations at the site in the 1960s a skull was found (variously described as a steer's skull and a horse's skull) with a dice placed on top of it showing the six uppermost, hinting at possible magic ritual usage.

One of the best known ghosts of old Luton Town was 'Old Cleggie' who haunted the Clegg and Holden printing firm in a solid Victorian house in Old Bedford Road. He was last seen in 1975 by a new employee who described him as a slim man about fifty with a sunken face and wearing a grey suit. An older employee recognized him as Mr Clegg, one of the founders of the firm, who had retired in 1950. Many strange happenings were reported over the years. A mirror suddenly smashed on its own after someone decided to clean it, so that another employee's remark became prophetic; he had said that Old Cleggie wouldn't like that... Jumpers and coats were thrown off their hooks to the other side of the room; a circular saw with a foot pedal drive that was stiff from disuse would often suddenly begin to move; an old rum bottle flew through the air in front of several workers and things would regularly, unaccountably go missing, only to turn up in unusual places. Staff had the uncanny feeling that they were being watched, as though the long departed boss was still checking on their work: despite the electricity being switched off at the mains, lights were seen at night when the building was empty. Clegg and Holden is no more and the place remained empty for some time afterwards...or did it? It seems that at long last 'Old Cleggie' finally accepted his retirement.

A haunted street was the subject of an article in The Luton News (6th April 1994): earlier the newspaper had carried an item about Eddie Herbert and Sam O'Reilly who believed their house in Crawley Road to be the focus of poltergeist activity. This story prompted 70 year old Mrs Eileen Lewis of Essex Close to contact the newspaper, to tell them her own story about an old haunted house she had played in as a child back in 1936. Now demolished, the house had also been in Crawley Road Luton. The twelve year old child responded to a dare from her friends to go to the top of the empty old house and look out from the window. 'I remember it in exact detail,' she told reporters. 'I went into the room and then all of a sudden stones were flying around me; I don't know where from they just came out of everywhere. I was so scared I ran down the stairs while the stones were all around me but I was not hit by them.' Is her story hard to believe? 'I will swear on a stack of bibles that what I'm saying is true,' she added;   the unfortunate Mrs Lewis would indeed appear to have encountered a poltergeist. Poltergeist 'infestation' is often typified by stone-throwing and unlike ghosts poltergeists prefer company, are noisy, mischievous and often 'perform' in daylight. Of the multitude of cases I have studied, as in this case, there is more often than not a young person involved. This leads me to believe that whatever a poltergeist may be, in many (though by no means all) circumstances it requires the particular energy of the young in order to function most effectively.

I had occasion to visit Cresta House on the corner of New Bedford Road and Alma Street in Luton a couple of years ago. Later on I learned that the old Alma Cinema used to stand on this very spot right up till 1960, when it was demolished. The Alma was a real 'picture palace' when it opened at the end of 1929; opulently decorated, as was the fashion of the day, it could accommodate over 1,600 patrons. A huge glittering globe gave out a powerful red light atop the frontage. On the floor above the cinema there was a restaurant with a ladies' orchestra and a ballroom. Even before it was built however, the site had the reputation of being cursed. Certainly there must have been resentment from the former residents as where the proud cinema now stood there had been terraced houses and cottages. Many people had suffered a forced upheaval when they suddenly had to find alternative accommodation. What is known for sure is that a workman died in a tragic accident during construction of the cinema, when he fell from the roof girders into the partly-completed circle.

Staff at the Alma often complained of an 'uneasy, oppressive atmosphere' at the back of the building. This seemed to reach all the way up from the boiler-room to the projection suite at the top of the theatre. The area was used solely by staff; it consisted of a maze of rooms reached by two concrete staircases with interconnecting rooms and passages at various levels. Going upstairs could be an ordeal as cinema workers felt a 'presence', always one flight ahead of them. Descending was even worse; they felt as if 'something' was following them intent on pushing them downstairs. The slow, steady decline in the cinema's fortunes could not be halted, from the end of the 1940s, despite attempts to introduce variety shows and stage shows as well as the movies. By the early 1950s the Alma had lost popularity to such an extent that it was converted to the Cresta ballroom; initially well-received, the ballroom also failed.

During demolition in 1960, in the circle girderwork, an old cap was found heavily encrusted with what was thought to be bloodstains. Some people speculated that this was the headwear of the workman who had died here over thirty years previously.

Cineworld in Luton has a ghostly reputation too. Staff and customers alike have claimed to have experienced strange noises, icy temperatures and fleeting shadows here. Co-manager Bharti Parmar was quoted as saying, 'When it was the Co-op, they used to say it was haunted, and there's been some strange goings-on in screen six since we opened. I ensure I've got someone with me when I go in there!'

Leagrave Road was well known as the home of one of Luton's largest employers – SKF at one time was a major player in the manufacture of ball bearings, but like many other industrial casualties it no longer plays a part in the town's economy. Back in the days when SKF was a big business the factory had a resident ghost; she had been a cook and appeared in white overalls and white boots (quite appropriate for haunting)! One of the witnesses to the 'visitations' was Mrs Dora Rouget, canteen stock controller, of Duncombe Close in the town; Dora had a brief vision of the cook in 1973, which left her 'very taken aback but not particularly frightened.' Shortly afterwards the canteen closed and the sightings ceased. Changing fortunes had previously affected the one-time booming hat-making industry that was synonymous with Luton, but after the Second World War it went into a sharp decline. The association with the town dated from the late 1600s when Luton was mass producing straw hats. During its peak (no pun intended) the hat trade was pulling in as much as £1.6 million for cheap bonnets in 1885. Those far off days are commemorated in the football club's nickname – 'The Hatters'. In the twenty-first century it was the turn of car manufacturing to leave Luton, with the closure of Vauxhall Motors (Owned by General Motors of America); the new Vauxhall Iron Works had originally moved its new motor works to the town back in 1905. Now the town is dependent on the expansion of its thriving International Airport

A couple living in a flat at Green Court, Hockwell Ring, were forced to call in an exorcist after suffering with a malevolent 'presence'. 'It' was described by Jennifer Davies as a strange figure dressed in a black cloak like a monk's habit, with a horribly scarred and disfigured face. 'It' caused pots and pans to fly through the air without falling, other things to go missing and one time the unquiet spirit stood by the baby's cot and shook it violently. At other times there was the sound of a hot frying pan flaring up; neighbours told them that a man was once badly burned in the flat and died there, his body not being discovered for five weeks. After the exorcism the phenomena ceased.

Luton Museum and Art Gallery in Wardown Park, was once a gentleman's residence which became the property of the council early in the twentieth century. During the First World War it was used as a military hospital and it was at this time that the building gained a reputation for being haunted by the ghost of a housekeeper. Convalescent soldiers and nursing staff alike were occasionally startled by the ghostly grey female figure; when the house became a museum and art gallery the hauntings seemed to be reduced to the sound of footsteps sometimes heard on the back staircase in an otherwise empty building. There was one final sighting recorded however; in 1971 two heating engineers working late one night in the cellar heard footsteps approaching them on the steps. One of them caught sight of a woman wearing a long dark dress with a bunch of keys hanging from her belt. As he watched, the figure turned and went back up the stairs out of view. The engineers later questioned the caretaker about the woman. He assured the pair that neither he nor anyone else had been down to the cellar. I was told that the museum was recently the subject of a paranormal investigation and that it continues to be a haunted site.

Jack Boutwood claimed a sighting of the 'Luton gasworks ghost' in 1961. He described it thus; 'She had jet black hair, was well over six feet tall, and aged about forty, with a white robe that hung down to her feet.' He had just left the gas works at two a.m. for a smoke, when he saw the large translucent figure of the woman gliding towards him. He quickly jumped on his motorcycle and drew level with the apparition for a better look. Returning to the gas works he told his friend Bert Fleckney, who managed to get a glimpse of the phantom as it disappeared around the corner into Highbury Road.

A pub in Leagrave Road called the Horse Shoe (now a hotel) had a strange history. During the 17th century a man named Cain was murdered in a nearby field; his corpse was taken to the Horse Shoe where the suspected murderer was asked to touch the body. When he did so blood poured from the wound and soaked the floor. For many years afterwards the bloodstains continued to reappear.

The strangest reported 'ghost' sighting in Luton took place in March 1969; on a narrow trackway leading from Canesworde Road to the Dunstable Downs two witnesses described a weird encounter. They were Martin Leach of Croft Green and Robert Wright of Chiltern Road and they gave a most detailed impression of the 'thing'. Martin's account was, 'It was white at first then it turned black; it had a big hat like a trilby, and glided about eighteen inches off the ground, it didn't have features like a person, it was about eight to ten feet tall and very broad.' Robert added, 'It didn't have any arms and legs but was wearing a large hat and it floated rather than moved.'

William Clifford's ghost is reputed to haunt the Four Horseshoes pub in Park Street. His inn was burned down in 1876 and his body was found by the back door together with a cash box holding the night's takings. Staff at the Four Horseshoes have reported hearing the sounds of money being counted when they were in the cellar and of suddenly feeling cold even on the warmest of days. Another Luton 'local' that has a reputation for being haunted is the Cork and Bull, where 'a vision in white', named 'Ann' has been credited with moving things around, shaking tables and throwing bar stools across the room. There are reports of her moving along the road from pub to shop and appearances on the same day in different locations. People have claimed to have sat and spoken to her, the response has been either that the ghost has touched them or something has been thrown across the room. There is a story that 'Ann' is the spirit of an old woman who lived here in times past, who was murdered in the pub...could it be that her restless spirit is trapped in The Cork and Bull?

Such is the current level of enthusiasm for all things paranormal that places with any hint of being haunted draw large numbers of people to the spot. One example is the site where Ellen Castle's dead body was discovered in August 1859 – Sundon's chalk pit. She was hacked to death with a knife by her husband Joseph, who then dumped the body at the pit. Members of the public with an interest in the supernatural collect at the site, particularly at Halloween, in hopes of seeing Ellen's wandering spirit. My experiences lead me to believe that these folk are more than likely to be disappointed, as ghosts tend to be shy and elusive. The more people that gather the less likely are the chances of seeing one, only poltergeists seem to prefer company.

Since updating this book I came across the following new story from Luton/Dunstable on Sunday, the April 4th 2004 edition. Grocer Nissal Ahmed and his wife Rizuana were too scared to live in their flat in Dorset Court, Bailey Street in Luton, so they moved in with Rizuana's parents in Marsh Farm. Their problems began when Rizuana felt uncomfortable in the flat but was unsure why. Then on the Thursday night previous to the report, Mr. Ahmed was watching television when an apparition, of a headless man appeared in the doorway for a matter of seconds. The next day he and his wife were talking in the porch when he felt a liquid splash onto his face. He said, 'I asked my wife what it was and she screamed. I looked in the mirror and realised it was blood. I checked myself and my wife in case we'd cut ourselves but I knew we hadn't and that's when the panic set in. I ran to the bathroom to wash it off and as I did, saw more splattering on the walls. It was a terrifying and there was absolutely no explanation for it. I rang the police but they said no crime had been committed and advised me to ring a priest at my mosque.'

The Ahmeds followed police advice and three days later a form of Islamic exorcism took place in the flat, carried out by Professor Masood Akhtar Hazarvi, a High Priest at Luton's Central Mosque, he was accompanied by Muslim community leader Khan Anwar. Professor Hazarvi said, 'When we went round there was blood on the walls. It was genuine. We recited verses from the Koran and also prayed and put readings on the walls where the blood was. I hope that if there were angry spirits in there they will have gone now.' Mr. Ahmed added, 'We have lived here for three years have always liked it and never wanted to move out. Now my wife has vowed she will never return. The previous person who lived in this flat was an old man who simply disappeared.' A neighbour said, 'He just vanished without explanation. No one knows whether he is alive or dead.' My thoughts on the case are as follows; was the old man who disappeared murdered and possibly decapitated, and was his spirit attempting to make contact to tell someone the truth? We shall never know, but it is a strange tale indeed...

The most interesting case for APIS in the Luton area was the BBC Radio Three Counties investigation. I was contacted by Catherine Lynch, a friend of mine who had worked on local newspapers and graduated to a job in broadcasting. To my great surprise Catherine told me that the radio station was haunted and the staff wanted the matter to be investigated. Intrigued, I organised a visit on Sunday 15th August 2004, in the company of two mediums and two dowser friends, with Catherine as our guide, at the building at 1a Hastings Street. I learned that we had been called in because there had been many strange occurrences, culminating in three odd incidents in the space of three weeks in 2004. Jenna, a staff member, reported seeing a lady waiting outside the reception area on the first floor, yet when she had walked around her desk to see what she wanted, the woman had completely disappeared and there was nowhere she could have gone. It is a secure area and visitors have to be 'buzzed in' by the receptionist, who unlocks the door electronically from her reception desk, or gets up and comes around the desk for a better look at the visitor, then opens the door by hand. This happened twice and on a third occasion the receptionist felt that someone was there even though she didn't see anyone. There was quite a good description to go on – a blonde woman in her early to mid forties, wearing a dress and a hat; she had been seen facing the secure door with her head cast down. The sightings had all been in daylight, from a distance of just a few feet through the glass panel at the side of the door. A colleague has heard the reception door close at night when nobody was about.

Keith and June Paull arrived first and I set off with them to record their findings. Keith soon located a 'haunting energy line' right opposite cubicle 1b, near the recording studio on the ground floor. He was unaware of the fact that one of the staff, Jena, had reported an uncanny sensation here, as though something had been following her. In the reception area I saw Keith physically knocked off balance by the strength of the haunting line he encountered behind the reception desk. I had not told him that this seemed to be a 'hot spot' in terms of the hauntings at BBC Radio Three Counties. As we moved through to the newsroom Keith surprised me with a new revelation – the haunting line stopped dead in the middle of the room, precisely in the centre of the sports reporter's desk. This was the first time that I'd ever heard of such a thing, usually haunting lines run right through entire buildings. Could this mean that the spirits only acknowledge the area up to this point and from here on no ancient buildings existed before? Does this represent a fairly modern extension to the building I conjectured?   Keith felt quite 'woozy' on the back staircase at ground floor level and he detected a really strong negative energy line here. We later learned that a few people have 'felt things' and even seen shadows out of the corner of their eye in this area.   In the basement Keith discovered that the haunting line follows one wall, Jenna has never liked being down in the place, which is used as a store room, claiming she 'sensed something which lies in front of the entrance door' close by a column.

I thought that it would be interesting to have two different psychic mediums on this preliminary investigation, to compare notes. First of all I went around with Christine Gould and her impressions came thick and fast, she described herself as 'winded' on entering the building, that there was a stagnation of energies and we all felt an unnatural heat. We were to find out a bit later the reason for this heat, and the stagnant energies meant that 'people would come and go – a high turnover in staff', also staff 'can't be real in their self-portrayal and may feel more vivacious and enthusiastic outside this place.' I noticed Catherine nodding her agreement with these statements. In the corridor leading to the studios Christine sensed 'a vortex of energy, a spider's web of energies.' There were many spirits in the basement who would be nocturnal visitors to other parts of the building from about 8pm onwards. The psychic felt that in the past many elderly people had resided here – could it have been a residential home or a hospice in times long past? There was a sense of deterioration rather than sickness. Christine then picked up an association with women, specifically nuns in grey habits.

Next we went to the basement and Christine was of the opinion that this building had covered many floors, there were a lot of trapped spirits but they were happy. There were vagrants among them, made homeless through no fault of their own; this was a haven for them. There were many bearded men who had problems with their jaws; the spirits had known that we were coming. There was still a lot of anger in the atmosphere, which radiated upward, and this was the cause of the unnatural heat we had noticed earlier on, 'the heat of anger from centuries ago.' She was later to add that she felt that the basement had, in the past, been a kind of subway, a walkway where murder and rape had occurred, and once there had been some kind of railway line which had run close to here.

On the ground floor the dubbing room prompted the psychic to venture that things were mysteriously moved about, and staff had the impression of something behind them here, both of which statements Catherine was able to confirm. Further along the corridor the pool room had a sad atmosphere to it and a link to a suicide, possibly a former member of staff... In this room I would expect to feel calm, as it is used by staff for recreation; Christine felt foreboding, the presence of a troubled soul. On the stairwells the psychic noted a feeling of confusion being prevalent.

I made the next circuit of the building with Juanita Puddifoot, another highly experienced psychic medium. She told me that a few days prior to our investigation she had been visited by the spirit of an Elizabethan gentleman, who had met a most unfortunate end, having been garrotted by two murderers. This spirit had appointed himself our 'unofficial guide'. There had also been a murder in the building we were investigating, he informed her, where someone had been knifed to death with a fatal stab through the heart. There was a great deal of emotional pain around as the result of this stabbing. While Juanita had been sitting waiting for me 'the Elizabethan' had returned to her and there were a number of spirits about and they all cried out 'murder, murder, murder.' This turned out to be a particularly interesting perception in the light of what Catherine Lynch was later going to tell us, but at this time we knew nothing about the site's previous history. Like Christine, Juanita felt bad energy which would affect people, in fact at least one person working here was adversely affected by it and she felt sure that a male radio host would suffer headaches and digestive problems as a result of this negative energy. Another strong impression was of a woman who appeared in an all-white energy form, she was completely unaware of the modern world and anyone passing her by, around the reception area and on the ground floor. There was also a man, 'unpleasant and constantly reliving both anger and terror.'

Juanita's inspection began upstairs on the second level, where she saw a spirit person standing directly in front of dowser Keith Paull, the medium felt that people would often want to get downstairs quickly from this 'activity spot', and would see things on the periphery of their vision because 'there are a mass of fragmented spirits here.' We made our way down to level one and Juanita quickly picked up, behind the reception desk, 'paper or mail being moved around by three children.' Catherine immediately confirmed that things were mysteriously moved about there. Juanita tuned in to two boys and a girl, spirits who were happy to be in communication with her, there was a lot of sorrow and sickness attached, some of which is associated with the children, but some is attached to others and the children don't want to be surrounded by sickness and unpleasant adults. In the nearby newsroom an interesting 'visitor' made himself plain to Juanita, a knight holding up a sword, and he was the guardian of the haunting-associated green energy line that Keith had discovered earlier. Juanita felt that there was some significance in the fact that she had received visions even before coming to the site, 'of lines crossing, this was before my visit.'

On the ground floor Juanita was aware of a yellow, snake-like undulating energy line in the corridor, in the pool room a feeling of sickness was detected and another 'visitor' – this time a 'Cavalier', who bowed and doffed his floppy hat in greeting but didn't want to communicate further. Juanita asked Keith's pendulum to confirm the Cavalier's presence and it duly obliged by ceasing to swing in an arc and beginning to rotate in a circle. Near the live recording studio the psychic picked up an impression of a white-faced woman floating along, surrounded by an aura of sorrow.

We didn't check out the garage but later learned that one of the BBC staff, who is psychically sensitive, had checked it, and detected the presence of 'Eric', a 52 year-old pimp who choked to death on his own vomit while in a drunken stupor. It appears that he is upset by all the people who hang around the area he occupies and in the cellar a remorseful spirit has been sensed, he seems to be looking for the woman he strangled.

When we were all reassembled for a debriefing Catherine dropped her bombshell – it was known that a woman had been stabbed to death by her husband in the vicinity some time in the 1950s, and her body cut up into pieces and taken to the cellar. This revelation was most startling because it had been deliberately withheld from our team, who not only pinpointed this site as being the place where a murder had been committed, but also correctly identified a knife as the murder weapon, surely an unlikely coincidence... I wonder if the unfortunate murder victim was a woman who had blonde hair, was in her 40s and favoured wearing hats?

Kevin Seddon, APIS's Archivist/Researcher, was able to confirm that the BBC opened Radio 3 Counties, in the Hastings Road building, in June 1985, and before that the site had been a Co-op in the 1900s, with flats above and cottages at the back. Our enquiries, as they say, are continuing...

Chapter 17

Beers, Wines...and Spirits

People often ask me why there are so many haunted hostelries. Firstly, as already mentioned in some earlier stories, before there were many hospitals inns were frequently used to take in the bodies of murder and accident victims. They also often had a 'laying out' room where bodies were kept prior to burial. People who die violent untimely deaths are more likely to provide 'unquiet spirits' and remain earthbound. Secondly, it may have something to do with the age of pubs, as after the church the pub is usually the oldest building in a village. Where there is a lot of history there are often lots of hauntings. Bedfordshire has more than its share of pubs with phantoms, I have investigated many of them yet undiscovered inn hauntings still seem to surface quite regularly. Many of these are worthy of serious research and Anglia Paranormal Investigation Society has a considerable number of Bedfordshire pubs on its forward vigil planner. The majority of well-known cases are presented here for your consideration.

Probably the most curious name for a pub I have come across in Bedfordshire was The Bullnose Bat (now renamed) The Square. Situated in Saint Paul's Square next to the Corn Exchange, this Bedford Town bar was the subject of extensive refurbishment about five years ago. A gruesome discovery was made when some thirty bodies were dug up on what was obviously the site of an ancient graveyard. I talked with Lynsey Henderson, the assistant manager, who remembered that the premises had once been an estate agents shop. She told me that she had clearly seen a 'flying bottle' move itself into a waste container behind the bar as though propelled by some invisible hand. This was also witnessed by Craig Longshaw, the barman, who claimed to have a phantom cat which appeared in his bedroom, despite the door and window being firmly shut... A previous manager had a puppy which would not go near a certain area by the exit door where some people have noticed a 'cold spot'.

The Five Bells at Cople is in complete contrast to the previous pub. The building dates from 1690 and the first licence was granted in 1729 to Eliza Smith. David Burns and his wife Christine took over as licensees in the summer of 2001 and I came to know them as good friends after our initial meeting in August, 2002. The pub is full of old beams and church pews with timbers that were believed to come from Elizabethan times; it is a common misconception that timbers from Elizabethan sailing ships were used in old buildings, but this is not the case. The supposed ghost of a 'sailor' is therefore possibly a fanciful interpretation. The ghost of a man has been seen in the corner by the fireplace, sitting on one of the pews. He has the appearance of a matelot though: descriptions always depict 'an old man of indeterminate age with a pigtail, wearing a hooped shirt, 'puttees' around his feet and smoking a clay pipe.' Kate Henderson, the previous landlady, saw him on several occasions during her twenty-three year tenancy. In January 2002 John Pilgrim of Three Counties Radio visited the pub while working on a feature about Cople village. He learned that a lady called Linda Archer had seen 'the sailor' as recently as sixteen months previously. Sightings do not seem to be too frequent and appear to be mainly confined to psychic mediums. They have been aware of his presence when in my company, but I have been unable to see him for myself. I did carry out some checks with Keith Paull, who has over thirty years experience as a dowser; we had two sessions at The Five Bells. Keith used his pendulum and mager disc and quickly located two 'presences' – the 'sailor' to the right of the fireplace, where I had been told he has been seen by several witnesses, and a new (to me, and to the landlord) spirit-person, this one a female on the other side of the fireplace. I later learned from Dave Burns that this room had served as a 'laying out' room where dead bodies had reposed in times gone by. On the second occasion Keith picked up what he referred to as a 'recording'. This is not the same as a ghost but is more of an impression of someone who had spent a lot of time in the pub in the 1960s. He had sat right at the corner of the bar, which has been in its present position since the 1920s.

The Hare & Hounds is situated in one of the prettiest villages in the county – Old Warden. I talked with the chef/proprietor; Jago Hart to follow up a rumour I had picked up that his inn was haunted. Although he had not seen anything himself he soon confirmed the rumours; he helped to run the pub with his partner, landlady Jane Hasler. The room at the far end of the pub, known as The Chapel seems to be the most affected area. Tobias, a medium, sensed 'a happy presence' here during a recent investigation. One evening when all the customers had gone home, Michelle, the barmaid, looked up to see a woman standing by the bar. When she next looked around the woman had disappeared. About two years previously, while renovations were taking place, a Spanish barman was staying at The Hare & Hounds one stormy night, when he was frightened by the ghost of a woman in an upstairs room, though he was alone in the pub at the time.

Another nautical ghost haunts The Bedford Arms at Toddington. PRS (The Phantom Realm Society) investigated it in the winter of 1999 using all the latest electronic equipment. They declared it haunted to the owners, Charles Wells Brewery. The landlords at the time were Richard and Nicky Bollen, who didn't see anything themselves but were well aware of the stories surrounding the pub, which has a long history of being haunted. The story was that a certain sea captain returned home one day to find his wife and child murdered; he was so overcome with grief that he hanged himself from a rafter in an upstairs room (now a bedroom). Regulars reported the following phenomena: a man in the fireplace, being tapped on the shoulder and the sound of phantom footsteps. A priest was called in to exorcize the place, but according to landlord Jamie Bent, 'The Captain' was still being sighted by members of his family and locals in both the bar and the bedroom in recent times.

Charles Wells is a large family-owned brewery in Bedford that owns some 250 pubs, not surprisingly it has at least fourteen which may be haunted. The Cross Keys at Pulloxhill where landlady Sheila Meads with her husband Peter has been working for nearly thirty years, was one of the most haunted according to the paranormal investigation carried out by the PRS in 1999. Customers had reported seeing a grey lady and a man in a 1930s brown suit. The man was recognized by regulars as a previous landlord who died at the pub. During the ghost-hunters' vigil several shadowy figures were seen strolling across the bar, also a strange incandescent blue light, which moved across the floor.

I decided to hold a vigil myself with the help of my friends at Anglia Paranormal Investigation Society, to update the findings of the Phantom Realm Society. So it was that on the damp and chilly night of Saturday 22nd November 2003, a team of eight APIS stalwarts set up base in the Small Bar at The Cross Keys in Pulloxhill. It was a night of the full moon and quite dark in the pub once all the lights were off. By the time the last customers had finally gone home Sheila and Peter Mead had retired upstairs to bed, and we had set up our recording equipment, it was 12.30am. Keith Paull and I 'mapped' the premises, and discovered a 'haunting power line' running at 265 degrees diagonally through the restaurant. He also identified four other 'sensitive spots'. Keith had been on the original investigation and had kept his notes; the restaurant 'power line' was nearly twice as strong as it had been in 1999. I then went around the building separately, with Joan Dancer, our rescue medium, who was unaware of Keith's findings, and asked her for her observations. I was deeply impressed when she correctly identified the same 'energy signals' in exactly the same places! During my 'walkabout' with first Keith and then Joan I noticed a most curious thing. The Meads' dog, a German Shepherd, avoided the invisible 'power line' in the restaurant and chose to creep around the edges of the room rather than walk directly the shortest way across the middle of the room, even in the dark. Also, although normally a friendly pet, the dog started growling as soon as Keith began using his pendulum to dowse. A subsequent photograph taken by Joan in the dark shows an unexplained point of light on the dog's nose.

Joan Dancer has been a Spiritualist for over thirty years and is the most experienced psychic in our group. She received numerous impressions throughout the night. While sitting in the small bar she felt the draught of movement as a lady went past on several occasions. She felt that this was Mary, who had three children, two girls – Jane and Meg, and a boy – William. Mary was 56 when she died in 1874. Joan also saw this same lady sitting sadly by the fireplace crocheting with a well-worn bone needle which moved extremely fast. In the 'Cabin' Joan felt a cold spot and a 'crackling energy' which I also witnessed, rather like static electricity, in one corner of the little room. There was also the presence of a soldier in a red uniform with a pillbox hat and highly polished shoes. He had fair hair with a parting on the right side and he was called Harry Atkins. Another impression was of someone called George, wearing a white crew-necked shirt and with a double set of buttons on his breeches, he was probably a farmhand. Another man came through; this one had drowned in a pond close by the pub.

Karl, who seems to be not at all psychic, momentarily saw two small unexplained points of light in the restaurant at table height. Keith too, saw a strange light but this was a red patch, which appeared fleetingly on the dining room wall. We all clearly heard the sounds of a baby crying each time we went to use our walkie-talkies, though there were no children staying in the pub. We would not claim that this was in any way paranormal as the sounds from baby monitors can be picked up over some distance, but it was strange how the sounds cut in every time we went to talk...

It would be useful if some of Joan's impressions could be checked out, e.g. is Harry Atkins buried somewhere near by, could Mary be traced from the few sketchy details that were picked up? We were able to establish that there had been a drowning in a nearby pond some time in the past.

The most interesting thing to emerge about The Cross Keys and the reason Joan had been so keen to attend our vigil, was an incident she had back in 1986. Joan had visited the pub all those years ago with her two psychic friends Norma and Valerie. They had been hoping for some paranormal impressions but it seemed to be a quiet day and Joan was just remarking 'Well this is a dead loss...' With that her teaspoon floated off her saucer and slowly descended into her lap as though lifted by invisible fingers. Fortunately it was witnessed by Valerie as Joan could not believe her eyes. It left a lasting impression with Joan so it was no wonder that she insisted on joining us on this occasion! Having worked with Joan for a good while now I can say that her testimony is particularly valuable and I trust her honesty and integrity one hundred per cent.

Bedford and Kempston Herald carried a report on The Falcon Inn in Bletsoe in June 1996. The young (26-year old) manager, Steve Elliott, was the victim of a playful poltergeist. The seventeenth century pub was first managed by Steve in October 1995. Almost immediately psychic phenomena began to plague him: a pair of large bookshelves fell from the wall then they began floating around the room and mysterious footsteps were heard. 'It's not just me who has heard the footsteps either, he reported, 'there were a few of us here one night when we heard them upstairs. We all rushed up to see who it was but no-one was there.' Keys have flown across the room and bottles have fallen from shelves. 'The business of throwing things around seems to happen more when I have a girlfriend here,' he explained. One time bottles, books and glasses flew straight at one of his girlfriends as he sat with her. Steve's belongings have been stolen by the ghost only to reappear a few days later. '"It's a spooky feeling,' he admitted. Initially nobody would believe the young manager but after more customers saw the spirit they were convinced. There may be more than one ghost as Steve has heard several voices whispering at the same time. Although he was beginning to get used to the friendly ghost he was often woken in the middle of the night thinking that there was an intruder in the place.

I decided to visit the former coaching inn and update myself on the hauntings at The Falcon. It was a fine, sunny, late August day in 2004, following weeks of interminable rain, when I finally got around to checking it out. Easily accessible right on the A6 and beautifully located close to the river with attractive gardens, this was my sort of pub. Lisa Sutcliffe, the manageress, confirmed that the site was still active. Glasses regularly crack for no apparent reason in the bar area close by the fireside and she knows of at least three occasions when glasses have been 'forced' off shelves to end up smashed on the floor. Lisa recently talked with a customer who had worked as a chef in the pub, and he was present, along with several other people, as a glass decanter containing cordial had its stopper lifted and the decanter was moved along the bar, seemingly by invisible hands. Waitresses in the restaurant have had strange experiences while alone in the room, one had something thrown at her and another heard her name called. Lisa is unafraid of the ghost and like Steve Elliott before her, feels that the 'presence' is a friendly one.

Not far from Bletsoe are Sharnbrook and The Fordham Arms, which I visited to meet Simon Harrison, the landlord, who had called APIS in to investigate his inn. While I drew up a plan of the pub's layout the young landlord guided me around his premises, which had once been a hotel and which has had two downstairs extensions. Visiting mediums from the north of England had declared that the site is home to no less than six spirits. It was only twelve months previously, Simon told me, when he had been downstairs in the dry cellar underneath the bar area by the ice machines, when a plastic scoop 'whizzed past my ear', he had definitely been alone in the cellar room at the time. Some six months prior to this, one Monday morning, just after he had taken a frozen food delivery, he had been in the bar by the back door and he had bent down to turn on the radio when he saw a shadowy figure, which he thought might have been wearing a cape, flit diagonally across the room and disappear through the wall where the blackboard hangs. He told me that there was once a doorway at this point but it had been filled in. I left after we had agreed a date for the APIS vigil.

I mustered an APIS team for the night of Saturday 2nd September at The Fordham Arms – there were seven of us – Vanessa, Sarah, Liz and Emma, Paul, Andy and myself, plus landlord Simon, who was an enthusiastic participant. If you are organising a vigil at a busy bar expect to start late – there are always late night revellers who don't want to go home early and we were not able to set up our equipment until 11.30pm. We didn't have to wait long for the most spectacular event of the night. Andy Garrett, our Investigations Co-ordinator who plans most of our vigils, gamely organised things so that Liz and himself had the first watch together in the darkest depths of the cellars. Andy is a most down to earth person but he is what psychics refer to as 'clairsentient', that is he 'feels' the presence of ghosts, he can detect them even when they remain unseen. It was proving to be an uneventful hour for Liz and Andy though, so towards the end Andy made a provocative remark to the effect that 'if there was anything there it had better make its presence known', with that he felt an unseen energy rushing straight through him, Liz felt the backlash in the form of 'a fierce wind', that fanned past her and straight out of the cellar. For Andy it was an extremely unpleasant and unnerving experience, but fortunately it was not the first time he has been through such an ordeal. He was able to calmly ask over the walkie-talkie, for the lights to be switched back on, but I could tell by his voice that he was shaken by his encounter, and who wouldn't be? The general consensus amongst the team at the end of the vigil was that the Fordham Arms deserves its reputation as a haunted pub.

There has been a pub on the site of The Cross at Beeston since 1605, indeed the oldest part of the present building dates back to 1700. In the spring of 1995 the landlords were Barry and Sylvia Haslett; Sylvia talked about her paranormal experiences there. 'Although we were warned by the previous owner about it (the ghost), we didn't believe it, but it has turned out to be true – there's no doubt about it.' A fairly recent incident involved some pool players. They were scoffing at the idea of a ghost when the cue chalk resting on the side of the table threw itself into the air and some time later a bottle of best whisky flew off the shelf and smashed onto the floor. Sylvia explained her attitude. 'I'm not frightened of the ghost and I think he's mischievous rather than malevolent. When I'm in the cellar I get the feeling he is watching me to make sure I do it properly.' Other happenings include the beer gas being turned off by itself and the lager beer line has sometimes been changed back to an empty barrel after Sylvia knew she switched to a full one. It all began shortly after the Hasletts moved in a few years previously. Sylvia continued, 'One Sunday night shortly after we took over the pub we thought we had offended people because no-one had come in for a drink. Then I heard a noise outside and realised customers were waiting in the car park because the door was bolted and locked against them. We hadn't touched the lock and the keys were hanging behind the bar.' 'On another occasion two ladies got trapped in the lavatory because the dead locks had been turned after they had gone in.'

A former landlord reputedly haunts the Oakley Arms at Harrold; Dan Orpin ran the pub in the late thirties and early forties. Since his death in the village he has appeared in front of drinkers in present times, either running through the pub or smoking a cigarette and having a pint. A quiet ghost this, his footsteps are sometimes heard upstairs but he seems to prefer sitting downstairs in the public bar, just enjoying a pint with the other regulars.

Northill is another picturesque Bedfordshire village and The Crown is perfectly located in the shadow of the church, surrounded by sprawling gardens and woodland. It has been attracting drinkers since 1780, when it became an inn – before that it was a vicarage! The pub's cellars are reputed to be the entrance to underground tunnels that ran between the pub, the church and Ickwell Bury, which was formerly a monastery connected with the Order of the Knights Hospitallers. Tony Dawson, a previous landlord of The Crown, said, 'Legend has it that one of these tunnels runs under the playground area and across the daffodil field at the back of the pub. Several of the customers have seen the figure of a monk appear from the Green beside the pond in front of the pub and glide across the car park on to the lawn, before disappearing into the woods.' When he was living above the pub (he was there for seventeen years) his sons had a ghostly visitor. 'Some years ago (around 1994) my sons encountered a figure in black coming towards them in the upstairs corridor one night. They both rushed back into their bedrooms to arm themselves with snooker cues but when they came out again the figure had gone.' They later discovered that a robed figure, presumably a monk, walks the corridors of the pub. Ian Taylor followed Tony Dawson as landlord of The Crown, and he admitted that certain areas of the pub had 'a presence' about them. 'There is an area behind the bar, that if you stand on it with bare feet, is always warm as though someone else has just been standing there when there has been no-one else about. The strange thing is there is absolutely no reason for that spot to be hot because it is not near any pipes, boilers or central heating.'

The Bell at Toddington is a haunted pub that has not received much publicity, so I went along to meet the landlord, Marcus Hayden, to find out about the rumours I'd heard, for myself. His resident ghost is one that makes its presence felt rather than appearing, with most of the 'activity' being upstairs. That Christmas would be Marcus' fourth at The Bell, formerly The Bell Inn, and the 'Inn' part was dropped because it seems to imply that a pub has rooms to let, which this pub doesn't. Three years ago a lot of refurbishment was carried out, something which seems to set off the ghostly activity, as with many other cases I have investigated. A room upstairs is known as 'the slopey room' on account of its angled floor, and woodworm was found in the floorboards, which had to be replaced before new carpet was fitted. The carpenter had left around four o'clock in the afternoon, when an 'almighty bang' came from the empty room. The door behind the bar was open so the noise was clearly heard by Marcus and soon after another loud bang rang out. The landlord rushed upstairs and checked the flat, found nothing and left the door open again, as he wanted to ventilate the room to clear the smell of the chemicals used in the woodworm treatment. What was particularly odd was that the door in question is a cottage-style one with a Suffolk latch which needs to be held down in order to close it; also the door naturally swings open, not shut. When, some five minutes later there was another huge bang, Marcus was in no mood for the spook's antics; he rushed upstairs once more and stood firmly in the middle of the room. He yelled at the invisible presence and told it to stop; the alterations were necessary, and he was only trying to improve the pub. It worked, because the door-banging ceased; despite the 'tantrums' Marcus feels that his ghost is a 'friendly presence'.

Tess, the assistant manager, has worked at The Bell for ten years, and told me that she had an uncomfortable night at the pub in November this year. Her husband gets up early, at five a.m. to go to work, and there was a lot of noise then, with creaking boards and doors banging in empty rooms, but nothing was seen. One of the back rooms is 'as cold as ice', it 'hits you as soon as you step in', so that even in summer this bedroom is cold – which was confirmed by Marcus. Downstairs, one of the bar staff, Hazel, has felt a ghostly hand on her shoulder, which she thought might be the spirit of an old landlord calling 'time'. The shoulder-tapping has also been felt in the 'slopey' room upstairs. Some customers have reported seeing the ghost of a woman sitting on a stool at the corner of the bar. Tess also told me that a former landlord, Les, had an odd experience one night when he suddenly felt his bed sheet pulled so tight that he couldn't move his arms. He told off the invisible entity, asked 'it' to leave him in peace, whereupon the pull on the sheet ceased.

Interestingly the stories I had first heard about The Bell before visiting it were not identical with those of either Tess or Marcus but complemented their experiences. Rumour had it that a young girl was imprisoned in one of the upstairs bedrooms in the past and that poltergeist activity, including glasses being broken, had occurred here. The Bell takes its name from the bell foundry that used to be situated in the once-bustling market town. This timbered pub is a former coaching inn and dates from at least the sixteenth century, possibly even earlier.

I mentioned The George Hotel at Silsoe, in an earlier chapter and it seems that it is still the focus of paranormal activity. As recently as 2003 unusual occurrences were reported, hotel chef Jason Towey was quoted in the local press, 'When I used to live in the pub I had a few strange experiences. One time I was in the restaurant having breakfast when I heard the kettle boiling. I went into the kitchen and found the kettle was hot – but it was unplugged at the wall. Another time I was going to sleep and heard scratching at the door – but it wasn't like an animal scratching it was very long and slow. I opened the door but there was nothing there. Apparently there are about five ghosts here.'

Medium Heidi Crumpton led an investigation at The George Hotel in March 2003, which claimed to have made contact with the spirits of Lady Elizabeth de Gray and William Yorke, the former is a well-known historical figure, the latter has, so far, not been traced. Heidi was in no doubt that The George is haunted, she felt something in the corner of the cellar and landlady Mrs Crumpton later confirmed that she had never liked the far left hand corner down there. 'Something' was also felt by the investigation team in room three, the manager's office and the attic. It turned out that these are the rooms that the staff have traditionally believed to be haunted.

I gave a talk about the paranormal, which was organised by Darren Flint of Mid Beds District Council Tourism Office and Charles Wells Brewery, at The Cross Keys, Pulloxhill, during National Pub Week in February 2004. There I met Caroline Flint, Darren's sister, landlady of The Guinea at Moggerhanger; I promised to follow up an interesting lead that she gave me about her pub. It was to be five months later, on the 17th July that I was able to fulfil my promise. Caroline told me that she met a psychic from Great Barford, called Keith, who had visited her pub for a drink and ended up encountering a different kind of spirit altogether. He had managed to answer a question that had bothered her for many years – who was the little boy who had haunted her since she was fourteen and had first started helping out at The Guinea? Clairvoyants in the past had sensed a child's presence around her and had told her that the boy would never follow Caroline home, but he was always around her in the workplace.

Keith, not my dowser friend, but the Great Barford psychic, dropped in one Thursday evening and was enjoying a quiet drink when he started to appear rather vacant and a bit distant, suddenly he put out his arm as if to stop somebody falling over, a 'somebody' invisible to the other people in the bar at the time! Keith was able to tell Caroline that he had seen the stumbling ghost of a five year old boy called Benjamin, who wore shorts with long socks. Ben lived in the mid-18th century, had been a bit naughty, liked to play tricks and he had not liked two or three previous landlord but was happy now because he thought The Guinea, under Caroline's management, was a 'fun place'. Ben had hated the cold and was always around the fire; Keith said that the child often still sits with Caroline by the fire. His legs hadn't worked properly and he died so young through a disease. It seems that the fireplace (complete with ghost) is the only feature left of the original Guinea, built in 1622. In 1963 this Charles Wells pub was demolished and completely rebuilt in the style we see today.

Caroline told me that about a week before my visit, on a busy Saturday night, a packed pub witnessed a decidedly paranormal event. An old 'stamp book' (a record of samples that inspectors had taken from a variety of the inn's alcoholic beverages) in a wood and glass frame, had flown over people's heads and broken on the floor. It had not fallen straight down as might have been expected, had it done so it would have broken several plates on the wall below it. Instead it shot out horizontally, as if thrown by an unknown hand. I speculated as I examined the book, which had belonged to the Phoenix Brewery and was dated 1887 to1895, if young Benjamin had been responsible and what, if anything, he had been trying to communicate to us by this action...

During the early part of 2004 I heard several reports from local people about the reputation of The Blacksmith's Arms, at Ravensden, for being haunted. The Blacksmith's Arms dates back many centuries, it used to be part of the Ravensden Estate owned by the Wytes family and was originally an unnamed 'beer house'. Records from the 1890's report that the Blacksmith's Arms premises were licensed to Higgins & Sons while the occupier was James Fensome, who also seems to have run The Case Is Altered, while his brother David was responsible for The Horse and Jockey. It was further reported that one of the brothers lost a son in a drowning incident, in a quarry on the estate.

In recent years a young couple who had been locking up one night in the empty Blacksmith's Arms heard doors banging shut and rushed out to find nobody but themselves in the building, and they had been at opposite ends of the place, so who had banged the doors? A team of twelve of my friends from Anglia Paranormal Investigation Society, led by Vice Chairman Paul Keech, carried out an investigation on Friday 18th June 2004. As it is a busy pub they started recording at midnight and it wasn't long before Andy and Joan, who were working in separate teams, both felt a push in the stomach by some invisible force. Joan later sensed the presence of a fair-haired child; she also felt a cold draught and heard a thud from a nearby table. Strange coloured lights were reported, variously green, red and orange, by Paul and Andy and Jenny and Keith. Nearly everyone in the team experienced sudden temperature drops during the night, including Shirley and Eddie Household, the landlords, who enthusiastically joined in the watches, and it was such a promising vigil that APIS will definitely be returning for a follow-up investigation.

Novice ghost-hunters are well advised to begin their research down at their own 'locals', as these are some of the most accessible and most frequently haunted of all public buildings in this, the most haunted country in the world.

Chapter 18

The Wilden Poltergeist

Earliest records show that Wilden village was owned by twenty-four 'socmen', that is holders of small estates. It is probably their individual dwellings that form the nuclei of the various 'wicks' or 'ends' still in existence – Smartwick, Sevick, South and East Ends and the centre of the village, once known as Church End; at least four more 'Ends' have disappeared. Wilden has always been an agricultural parish, there are traces of terrace cultivation and it is possible that some modern farmhouses are built on older, Danish sites. Of the many estates Wilden Manor would have been the largest; in the Domesday Book it was assessed at a value of five hides and was owned by the Bishop of Bayeaux. On his death in 1097 the overlordship passed to the Crown, then the estate passed through several hands, including the St. Remy family at some time in the 1100s. Records mention Robert de St. Remy in 1164 and William de St. Remy in 1204. On William's death in 1224 ownership passed to his two daughters, Agnes and Elena. Agnes married Ralph Ridel and Elena married John de Pabenham and the estate was, for a time, divided between the two families. It eventually became solely John de Pabenham's property; before eventually coming into the hands of the Lucy family, until Sir Thomas Lucy alienated it to Thomas Rolt in 1569. The Manor remained in the Rolt family for nearly 200 years and then another Thomas Rolt sold it in 1732, to Sarah, Duchess of Marlborough. The Duchess died in 1744, at which time Wilden Manor was purchased by the Duke of Bedford, who held it until 1837, when it was purchased, together with the advowson (the 'patronage' of an ecclesiastical house or office; the right of presentation to a benefice or living) by the Reverend William Shove Chalk, two years after he became Rector of the church of St. Nicholas at Wilden. The

Chalks settled in as 'squarsons' (a mixture of country squire and parson) for the best part of another century, and it is this family that seem to be connected with the poltergeist manifestations at Wilden Manor Farm. The rating valuation of 1925 lists John Wootton as owner and occupier of Wilden Manor Farm; the valuer's notes mention 'lots of thatch, house nice, and buildings fair', the property extended to 142 acres. The present day owner is a Mr. Pell.

Wilden is not an abbreviation of Wilderness, as some people have believed, but means 'Willow Valley' and it lies along the South Brook, which rises near Keysoe and joins the River Ouse near Wyboston. Older inhabitants have been known to quote the jingle, 'Wicked' Wilden, 'Rotten' Renhold and 'Wretched' Ravensden.

The area was notorious as the happy haunting ground of Mother Sutton, an exceedingly evil Bedfordshire witch who lived in the neighbourhood and was hanged, along with her daughter (who was presumably found guilty of being a witch too) in 1612. Local legend has it that a bride and groom saw the hag as they passed through Wilden about seventy years ago. She was a terrifying vision – beneath her black bonnet her features were coarse and masculine, and as she suddenly turned her head the unfortunate bride was subjected to a malevolent glare. The incident cast a shadow over the wedding and badly affected the bride on her subsequent honeymoon. A well-documented account was the much earlier Goodall sighting of 1873. Mrs. Goodall and her daughter had an identical experience while out driving their pony carriage; they saw a figure in long black trailing garments seemingly gliding rather than walking along the grass verge, with the 'fiendish' expression and sudden disappearance as they drew level. Mother Sutton earned her modest living as a hog-keeper, before she quarrelled with a local farmer and allegedly cast a spell over his pigs. His pig-sty became 'a turmoil of hysterical squealing animals, fighting and rending each other's guts.' This was enough to send Old Mother Sutton and her daughter to the gallows. The last reported sighting on the road between Ravensden and Wilden was in 1973, when a man related his meeting with the witch in similar terms to Andrew Green, a well-known ghost-hunter of the time. The phantom was to eventually receive a Certificate of Authenticity from the Society for Psychical Research.

Bedfordshire certainly had a number of other witches – including Elizabeth Ocle of Pulloxhill, Alice Smallwood of Blunham and Elizabeth Pratt of Dunstable. In the 1660s and 1670s Elizabeth Pratt, who admitted to meeting with the Devil and concluding a pact with him on Dunstable Downs, was accused of bewitching several children, as two died and two others caught a strange disease after visiting Mrs. Pratt, who died in Bedford Gaol (she was there at the same time as John Bunyan) before her trial. In the atmosphere of fear, superstition and mistrust that was prevalent in those times, being an old woman and living alone was enough to mark someone out as a witch. Ignorant superstition meant that the uneducated masses really believed that witches had a contract with the devil, who rewarded their work with a healthy and wealthy life on earth. They were supposedly capable of performing actions that required more than human strength; they could change the course of nature and inflict pain on those who offended them. Outlandish tests were dreamed up to prove a witch's guilt or innocence: she could be weighed against a bible and pricked (as witches do not bleed) and tried by swimming her (if she floated she was a witch and was put to death, if she sunk she died anyway).

Wilden today is an attractive village with a brook, the remains of a moat, a small green and old houses and cottages gathered around the church – but does it conceal a dark secret from its past? There used to be six pubs in Wilden; today there is just the Victoria Arms, opposite the Manor, and poltergeist activity has even reached across to the pub from the Manor; as I was to learn first-hand when I later interviewed a number of witnesses. The village boasts many listed buildings, among them Crowhill Farmhouse, East End Farm House, Lion Field Farm House, Church End Farmhouse and Granary, Tudor Cottage and Village Farm. Seventeenth century Manor Farm House is a Grade II listed building. In 1995 a survey of 1,300 buildings and structures listed for special architectural or historic interest mentioned the Manor. Historic Buildings at Risk in Bedford Borough 1995 described it thus, 'totally derelict, no maintenance over many decades, partial collapse, serious movement.' It was reported to be a timber-framed 17th century farmhouse, within a working farm complex. An appeal against a Compulsory Purchase Order was allowed in 1982 on the grounds that 'the building had been neglected for so long that repair was no longer practicable.'

I knew about the stories of the Wilden Witch, but not about haunted Manor Farm; which was brought to my attention by a young lady called Sheryn Firman. I interviewed her at her smart flat in Olivier Court, Bedford, together with her friends, Selina Hammer, Thomas Nilsson and Mark Dalrymple. It was an extraordinary account. Just a few weeks prior to our meeting the apparition of a woman had been seen going into derelict Wilden Manor Farm House. Screams have been heard on the site – a horse rider was allegedly killed after falling into a ditch in the village. All of those present swore that they had witnessed stones being thrown at them at Wilden on several occasions, even as far as the garden of the Victoria Arms pub, across the road from the Manor! Thomas Nilsson in particular, seemed to be most affected by the phenomena, having been to the site on numerous occasions. His car still retains a chip on its roof where a stone, flung from the Manor, hit his car as he left the village one night. A rock had landed next to him in the churchyard of St. Nicholas just the previous week-end. He felt uneasy about the place and sensed evil, yet was strangely drawn to this atmospheric site. His experiences included unaccountable smells – lavender and a smell of burning, rather like hair being singed. The most frightening aspect, however, was the poltergeist's ability to follow curious, enquiring people back to their homes! On one occasion, after having several stones flung at him on the track, Thomas had felt eerie both there and at home. He woke up at midnight to hear the sound of coins being rattled in invisible hands and yelled out at 'it', 'Leave me alone!' That seemed to do the trick as he wasn't disturbed any more that night.

An old lady in Wilden had told Thomas's friend Tony Toth that according to her grandmother a witch had been burned to death on the track leading up to the Manor, this track seems to be the focus of the ghostly phenomena. If so this would have been unusual but by no means unique – the majority of witches were either hanged or drowned. Tony knows more about the Wilden poltergeist than anybody, as he has had numerous encounters with the paranormal in the village where he went to school. There was a totally inaccurate story that at some time in the past the house had suffered a lightning strike which had caused a fire and that during this fire some people were killed, but this simply did not happen. One group of friends, who ventured onto the farmland surrounding the Manor late one night, heard ghostly talking and screams. There are a couple of dilapidated caravans on the track and local legend has it that two gypsy brothers used to live in them. One is supposed to have died of cancer and his brother died the following week – of causes unknown. Another story has it that the house was cursed by an unknown lady. It certainly seemed most promising for a follow-up investigation with such a reputation surrounding it...

On a foggy night in late November I decided to take up the challenge of investigating Wilden Manor for myself. It was evening when I met up with Tony Toth. He was then living in a terraced Victorian house at George Street in Bedford (he has since moved home). Tony was the first Toth I had ever come across, but apparently in his native Hungary the surname is as common as Smith is here. Tony is an interesting personality, good company and passionate about the supernatural. After a general chat we set off, and as it is a dark road out to Wilden I was glad of Tony's local knowledge.

We parked up beyond the Victoria pub and set off on foot to the church of Saint Nicholas. It was most atmospheric on this night, with its dark and foggy conditions, just a distant street lamp spreading its diffused light across the graveyard, like a scene from the film The Exorcist. The church itself is an imposing structure in the Gothic style; with chancel and five-belled square tower. Tony led me to a far corner of the graveyard that contained seven graves, quite separate from the rest of the cemetery. Where most of the area was well maintained this corner looked distinctly neglected and untidy. What struck me as distinctly odd was that all other graves' headstones faced the visitor, whereas the seven isolated graves had their inscriptions on the reverse side of the headstones. It is a mournful spot, surrounded by a group of trees that stand like silent guardians. We made our way behind the monuments to read them. The seven were all members of the Chalk family, once an influential village family. The stone is fading with the passage of time but most of the carving is legible. The seven are: Richard Gregory Chalk, a former Rector of St. Nicholas (in the mid-nineteenth century), Richard Landon Chalk (died 1863), Julie Chalk (died 1907), Mary Eliser Chalk (died 1918), Thomas Craddock Chalk (died 1950), his wife Annie (date of death illegible) and Gladis Mary Chalk (died 1954). I later checked the church records and found listed there three Rectors of St. Nicholas by the name of Chalk. They were William Shove Chalk (1835), Richard Gregory Chalk (1849), and Richard C. Chalk (1941). Why weren't they all buried in the same area of the churchyard? There was no sign of either William Shove Chalk (Richard's immediate predecessor) or Richard C. Chalk's (the last Chalk Rector) graves. The Post Office Directory for Bedfordshire of 1864 provided some interesting information too. 'The living (of Wilden) is a rectory, in the patronage of Mrs. Eliza Chalk, annual value £400, with residence; the Rev. Richard Gregory Chalk, M.A., of Trinity College, Oxford, is the incumbent.' I didn't experience anything untoward while we were in the cemetery but Tony twice got a brief aroma of lavender, a phenomenon he has regularly experienced in Wilden.

I decided to pay another visit to the cemetery the following month, this time in the company of Julie, my medium friend, her husband Paul and Thomas Nilsson. Julie's first words when we approached the Chalk graves were, 'There's been a cover-up', an interesting remark – might there have been a skeleton in the Chalk family's cupboard? Soon all the others were getting strong wafts of lavender, but not me. Julie was receiving more impressions, and beckoned me over to two gravestones, probably those of Thomas and Annie Chalk. When I put my hand, as instructed, on one of the headstones, I immediately got a strong smell of lavender, and continued to get occasional wafts thereafter. I wondered what the significance of this ghostly aroma could possibly be. Julie was under the impression that one of the Chalk Rectors had always had lots of it around the house, and she felt it could have been used to ward off evil.

Lavender has always been a symbol of purity; it has a history that stretches back to ancient Grecian times. The herb is also associated with Englishness and reached a peak of popularity in Victorian times. It would have been readily available to one of the Victorian Chalk family Rectors, as it was grown in quantity in the neighbouring county of Hertfordshire, at Hitchin. Queen Victoria is said to have had great faith in it, and it was probably used as a domestic disinfectant in her households. Long ago lavender was used for a wide variety of healing uses, from snake-bites to liver disorders. The sweet-smelling herb has also been credited with occult powers; it was one of the herbs consecrated to Hecate, Grecian goddess of infernal regions, who presided over magic and enchantment, and was patroness of witches. Lavender was persistently sought by witches, who could appreciate its value and knew how to transform innocuous plants into deadly ones.

Tony and I continued our investigations at Wilden and moved on up the road, passing The Victoria pub, to the Manor entrance, which is diagonally opposite the inn. There is a white gate that leads onto the property, guarded by a huge and ancient tree. Some people feel distinctly uncomfortable around this towering sentinel, but I merely found it impressive and wondered what its age might be. We made our way up the grassy track, which gently curves up to the ancient house. There was just enough light for us to pick our way along. We crossed a muddy pathway and as we approached the building it seemed to get colder. The Manor was a little disappointing, for it is a small building, cosy rather than grand, with just four bedrooms. Part of the roof had recently caved in but as I later learned, it had nothing to do with lightning, as Tony had been led to believe. The old house is crumbling, with only a few panes of glass left in the windows and some extremely solid oak beams are exposed. We had a look at the darkened interior with the aid of my powerful mini torch. Old tyres and general debris have been dumped inside, so we made our way carefully. The stairs are gone and the first floor is reached by an ancient wooden ladder. Upstairs the ceilings are bowed and ready to collapse at any time. The elements have taken their toll. The bathroom is open to the stars due to the roof collapse. I certainly didn't pick up any bad feelings in the place; although empty, the house had a peaceful air to it. Tony pointed out that he has not had any bad experiences here.

We made our way from the house back down the grassy track. It was a different story outside. The first oddity was the sudden noise of goats. We had been unaware of their presence on the way to the Manor, all had been extremely quiet. As we passed the caravan that had been a home to two gypsy brothers, there was a bang, as a stone was hurled at the caravan. We literally stopped dead in our tracks and around us was absolute silence. We were alone in the grounds, so who had thrown the stone? After pausing to look and listen we made our way to the ghostly white shapes of the grazing goats. They neither moved away nor made any sound, they seemed untroubled by our appearance, and so what had spooked them earlier? We returned to the track and were puzzled by a new phenomenon. A single tree is sited opposite the derelict caravan on the other side of the track. It is surrounded by scrub. From this area came a steady pattering noise. There was something I didn't like about the tree and its surrounds. It wasn't raining but the pattering continued. Tony saw small pebbles falling amongst the undergrowth and I could distinctly hear dripping noises and see small movements in the tree and scrub. It was decidedly eerie. Then another bang sounded from the caravan behind us, as if another missile had been flung with some force. This was all the evidence I needed of spontaneous poltergeist phenomena. Tony was delighted that I was witness to something he has heard on countless occasions. He also picked up a smell of burning, but I couldn't detect it – again it is something that frequently happens to him here, often on the track, which seems to me to be the focus of the poltergeist's activities.

We decided to go to the sheds at the far end of the property, to look for signs of the old well, reputed to be on the site. We were making our way over, slightly hampered by the mud, when we heard a terrific clattering on the corrugated roofing, as though handfuls of stones had been thrown up there. We reached the sheds, which shelter worn out old tractors, and both listened to a steady pattering noise on the roofs, like a steady rainfall, yet there was no rain. There was no sign of any well, as it was probably filled in years ago. We both felt cold back in the near neighbourhood of the house, but it was not a particularly cold night. By the track a chilly wind blew, giving me a decidedly cold neck. Once again we made our way back towards the main road via our trackway. Tony told me that the stone-throwing always happened on leaving the house, never on the way to it. At first I had thought that we had an angry poltergeist, warning us off, guarding its territory, now I was not so sure. Tony felt that whatever 'it' was, wanted us to stay, not go, and was drawing our attention to it. 'It' wasn't finished with us yet. We were nearly off the grounds now and Tony showed me where he had witnessed the most amazing phenomenon, a phantom fire, a bit bigger than a human figure, which lasted for about five seconds duration. This has also been seen by several other witnesses, I was later to interview one of them, Paul Huckle.

As I talked with Tony an even louder bang took us completely unawares; it was as though a larger stone was thrown in our direction, and it clanged off some long-abandoned farming equipment, right beside us. Then we noticed more pattering noises, this time coming from the huge tree by the gate. Tony felt a small tap on his shoulder, as though a pebble had landed on him from the overhanging tree. I heard a loud 'plop' at the same time. We made our way out of the grounds and noticed that it definitely wasn't as cold out here; the fog was much lighter too. It was around midnight, we had been at the Manor for over two hours, and I was completely satisfied with a most worthwhile night's investigations.

After my experiences at Wilden I was eager to see if other witnesses were prepared to talk about haunted Manor Farm House. It wasn't long before I met Paul Huckle, who had experienced quite a lot of poltergeist phenomena over several years. He had returned to his (locked) car at Wilden to discover a pile of bracken on his front seat and stones on his back seat. On starting the car, his radio, which had been switched off, came on full blast. He had heard a little girl's high-pitched scream in the middle of the night, in the locality of the Manor Farm House. At two a.m. he has seen lights on in St. Nicholas church, although it was empty. While standing at the gates to the farm in broad daylight he has seen stones, at first hovering in mid-air, then being hurled by some invisible force towards him, although they never seem to do anyone any harm. This was a particularly exciting report because it is extremely rare, in my experience, for any witness to observe hovering objects before they strike, usually the first warning of poltergeist phenomena is the noise of the object hitting something. This was the first and only time so far that I have actually met a witness to such a rare, almost unique, sighting. When picked up, these missiles, more often than not stones, are found to be warm to the touch. The most spectacular sight on the track, however, was the phantom fire that sprang up about five feet from the gate. Paul told me, 'It started as a small spark and flared up to some two feet tall, at which point I decided to get out of there.' Tony Toth was with him as we spoke and had also been present when Paul saw the anomalous fire, he added to the description, 'It kept growing until it was eventually about six feet tall by six feet wide. My friends made a run for it but I was just overawed, it's such an amazing sight. I have seen the fire apparition on three occasions, over a period of about ten years, so it's a rare sighting, and each time I am completely blown away by it.' He was right to be so impressed, for this, too, was a most uncommon phenomenon.

Tony had another story to tell, this one was quite amusing, it concerned the night when he had needed a taxi to take him home from Wilden. The taxi arrived and waited for him outside the farm gate. When Tony got in, the driver looked panic-stricken, called his base and told them 'Never again send me to Wilden'. On the way back to Bedford, Tony enquired as to what had bothered the driver, to be told that the taxi's windscreen wipers had gone on, although the ignition was switched off, also the petrol gauge had begun to flick up and down its dial, which thoroughly spooked the poor man.

Worryingly the poltergeist has also attached itself to Paul and Tony, as well as to Thomas Nilsson. Paul has had coins thrown around in his home after visiting Wilden, and drawing pins have unaccountably appeared both in his front room and in his car. He has tried to find some explanation for the weird happenings at Wilden Manor Farm, and decided to use an ouija board (something I always strongly advise against) to get some answers. Some messages were spelled out, to the effect that there was a fire, caused by lightning, and three members of the same family died, an old lady and three children, the names Emily and Fay and another that Paul couldn't recall, came through. I wondered if this séance, instead of contacting any spirits, had tapped into the sitters' unconscious minds, because they believed that a lightning strike had set fire to the Manor House and damaged the roof, whereas, in reality the roof was hit by a tree, but more of that later. Paul's son, Wayne, was able to confirm some of these incidents and to add more of his own: he has been at the farm gate and has seen three spheres of light dancing about a couple of metres off the ground. He was also at the farm with Tony Toth, and when Tony raised his arms in the air 'something' behind him mimicked his actions. It was a white form, just a torso and arms with no head or legs! Tony has had stones thrown around at his home on a few occasions, after night-time visits to Wilden. Tony seems to be somewhat obsessed with Wilden, which led me to wonder if he is the 'focus' for the poltergeist – they seem to need each other.

The whole affair intrigued me but I could only put forward theories as to what it all meant. One theory was that the entity of a little girl may have attached herself to Tony, seeking help by attracting his attention with stone throwing, so that every time he goes to Wilden Manor some kind of manifestation is almost inevitable. She would probably belong to the Victorian era, and she could have had a connection with the Chalk family, or even have been a member of it. Something terrible may have happened to her; some kind of abuse might have been covered up, an easy achievement for someone of local importance, influence and wealth, and she may have died in traumatic circumstances. Was her name Emily or Fay, and did her restless spirit need help in being finally laid to rest?

I returned to Wilden a couple of times after completing the first edition of this book, to try and answer some of those questions. I met Mr Pell, the owner, he knew nothing about the poltergeist activity but he was able to correct one popular myth, the roof had not been struck by lightning, more prosaically it had been damaged by a falling elm tree, which itself had been the victim of Dutch elm disease. On another occasion I brought along my trusted friend, rescue medium Joan Dancer. During the day that we visited the site she could not connect with any uneasy spirits and did not sense anything untoward about the place. This led me to reconsider my second theory – that the strange occurrences at the farm were all the product of Tony Toth's subconscious mind, he himself was responsible for psychokinetic (the supposed ability to move objects by mental effort alone) activity at Wilden Manor. Research on the subject of psychokinetics (commonly abbreviated to PK) has not come up with any conclusive explanation for this power, but some observations have been made, for instance that PK seems to be both a mental and a physical force at one and the same time. It has also been associated with psychological conditions such as neurosis, where acute anxiety may be converted into noise and movement of objects. But where, in all this, did the phantom smells and the fire apparitions fit in, and what about the times that Tony wasn't present, as when Thomas Nilsson was followed home by the phantom coin-rattler? There seemed to be no single theory which would neatly fit all the facts and perhaps that in itself is a recurring feature in all poltergeist cases...

I hope that one day we may learn more about the multitude of manifestations that have been experienced and attested to at Wilden by so many different people. I shall always remember it fondly as my first close encounter with a poltergeist, but I trust not my last! The poltergeist is both the most mysterious and most fascinating of any type of paranormal phenomena; it is also the most spectacular, producing incredible effects that can include raps and electrical disturbances, furniture-moving and the production of water puddles, stone-throwing and fire-starting. It is a formidably challenging enigma and I have every intention of investigating further and learning as much about it as possible in the future. Personally, the Wilden Poltergeist has been the most interesting, involving and well-authenticated modern ghostly phenomenon that I have encountered up till now in the whole of Bedfordshire.

BIBLIOGRAPHY

Fuller, John G.   The Airmen Who Would Not Die.   Souvenir Press 1979

Garbett, Peter J.   Bread.   Peter J. Garbett 1988

Gray, Adrian   Tales of Old Bedfordshire.   Countryside Books 1991

Haining, Peter   A Dictionary of Ghosts.   Robert Hale Ltd. 1982

Maple, Eric   Supernatural England.   Robert Hale Ltd. 1977

Puttick, Betty   Ghosts of Bedfordshire.   Countryside Books 1991

Robinson, Chris and Boot,
                   Andy Dream Detective.   Little, Brown & Co. 1966

Sutherland, Jonathan
          Ghosts of Great Britain and Ireland.   Breedon Books
                                                  Publishing 2001

Underwood, Peter   The A-Z of British Ghosts.   Chancellor Press 1971

Afterword

Little did I realise, when first captivated by Jenny Butterworth's experiences at Fairfield, what an incredible journey I was embarking on, in my exploration of the paranormal. It was to lead me in many new directions, where I learned many new facts, made many new friends and began many new adventures. It was to prove, quite literally, a life-changing experience. As I write, my research continues, on my next book, Ghostly Hertfordshire, which could turn out to be an even bigger and better book than Ghostly Bedfordshire. While I am working on that project there are several other books in the planning stage, so my future certainly lies in the pursuit of the paranormal.

During the course of writing this book on hauntings I joined ASSAP, the Association for the Scientific Study of Anomalous Phenomena and progressed with them to become an ASSAP Accredited Investigator. The Association was founded by one of my heroes – the late, great comedian and writer on the supernatural, Michael Bentine. My experiences with them fired me with such enthusiasm that I founded my own, local group – APIS, the Anglia Paranormal Investigation Society. What began as a handful of like-minded people gathering together has now grown to become the largest, fastest-growing and most successful paranormal group in the region. We have forty-five members currently (and still growing) of all ages (from 21 to 77) all professions and all creeds. Members come from Bedfordshire, Buckinghamshire, Cambridgeshire, Hertfordshire and Northamptonshire to attend our monthly meetings at The Five Bells pub in Cople, near Bedford. A second branch has been running in Hertfordshire for some time now and it meets at Baldock Snooker Club. There are, however, very real dangers in encounters with the paranormal without sufficient knowledge. Our watchword at APIS is all about making things safe for our investigators and teaching the right way to conduct investigations, and that is why we have a code of conduct. We have enjoyed some exciting vigils together – but that is for another book perhaps...

The ghosts that we encounter at APIS tend to be unseen – hence our motto – 'Let us find what you cannot see.' This is typical in ghost-hunting, most entities are experienced in other ways and actual sightings are relatively rare. The vast majority of encounters are heard (e.g. footsteps), smelt (e.g. roses, lavender or tobacco, most often), or felt (e.g. 'coldness – but sometimes heat). The mysterious movement of objects may be observed, a feeling of terror experienced, strange lights or shadows may be noticed (all of the aforementioned have been reported during APIS vigils). To help us record our findings we use the very latest equipment. Our Technical Consultant, Nigel Brockwell, can supply us with anything, from EMF meters to Ultrasound detectors, but they are a back up to our five senses (or in the case of our psychic mediums, six senses). Our ranks include a fourth generation dowser, Keith Paull, and a rescue medium with over thirty years experience, Joan Dancer. On the equipment side three most essential pieces of kit for professionally conducted ghost hunts are night-vision camcorders, EMF meters and digital tape recorders.

This is the Reinvestigated edition of Ghostly Bedfordshire, which means it is the second edition, with many new stories and illustrations as well as three entirely new chapters, on Dunstable, Biggleswade and Leighton Buzzard, which were not featured in the original. I hope I have made up for these shortcomings and done justice to those parts of the county which were under-represented the first time around! It has almost been like writing a whole new book, and just as much fun as before. Thanks for joining in the adventure.

AUTHOR'S NOTE:

If you have a true-life ghost story that you want to share with me (from any part of the British Isles) I will be interested to hear from you. You may contact me by email at the following address:

apismail@ntlworld.com